T0183483

Modern Birkhäuser Classics

Many of the original research and survey monographs in pure and applied mathematics, as well as textbooks, published by Birkhäuser in recent decades have been groundbreaking and have come to be regarded as foundational to the subject. Through the MBC Series, a select number of these modern classics, entirely uncorrected, are being re-released in paperback (and as eBooks) to ensure that these treasures remain accessible to new generations of students, scholars, and researchers.

Juan J. Morales Ruiz

Differential Galois Theory and Non-Integrability of Hamiltonian Systems

Reprint of the 1999 Edition

 Birkhäuser

Juan J. Morales Ruiz
Escuela Superior de Ingenieros de Caminos
Canales y Puertos
Universidad Politécnica de Madrid
Madrid, Spain

ISBN 978-3-0348-0720-3   ISBN 978-3-0348-0723-4   (eBook)
DOI 10.1007/978-3-0348-0723-4
Springer Basel Heidelberg New York Dordrecht London

Library of Congress Control Number: 2013953609

Mathematics Subject Classification (2010): 12H05, 37J30, 12-02, 32G34, 34C45, 34M15, 34M99, 34Cxx, 34-02, 37J10, 37-02, 58-XX, 70F07, 70H07

© Springer Basel 1999
Reprint of the 1st edition 1999 by Birkhäuser Verlag, Switzerland
Originally published as volume 179 in the Progress in Mathematics series

Cover design: deblik, Berlin

Printed on acid-free paper

Springer Basel is part of Springer Science+Business Media
(www.birkhauser-science.com)

Ferran Sunyer i Balaguer 1912–1967

* * *

This book has been awarded the Ferran Sunyer i Balaguer 1998 prize.

Each year, in honor of the memory of Ferran Sunyer i Balaguer, the Institut d'Estudis Catalans awards an international research prize for a mathematical monograph of expository nature. The prize-winning monographs are published in this series. Details about the prize can be found at

`http://crm.es/info/ffsb.htm`

Previous winners include

- *Alexander Lubotzky*
  Discrete Groups, Expanding
  Graphs and Invariant Measures
  (vol. 125)
- *Klaus Schmidt*
  Dynamical Systems of Algebraic
  Origin (vol. 128)

- *M. Ram Murty & V. Kumar Murty*
  Non-vanishing of L-functions
  and Applications (vol. 157)
- *A. Böttcher & Yu. I. Karlovich*
  Carleson Curves, Muckenhoupt
  Weights, and Toeplitz Operators
  (vol. 154)

Ferran Sunyer i Balaguer 1912–1967

Born in Figueras (Gerona) with an almost fully incapacitating physical disability, Ferran Sunyer i Balaguer was confined for all his life to a wheelchair he could not move himself, and was thus constantly dependent on the care of others. His father died when Ferran was two years old, leaving his mother, Angela Balaguer, alone with the heavy burden of nursing her son. They subsequently moved in with Ferran's maternal grandmother and his cousins Maria, Angeles, and Ferran. Later, this exemplary family, which provided the environment of overflowing kindness in which our famous mathematician grew up, moved to Barcelona.

As the physician thought it advisable to keep the sickly boy away from all sorts of possible strain, such as education and teachers, Ferran was left with the option to learn either by himself or through his mother's lessons which, thanks to her love and understanding, were considered harmless to his health. Without a doubt, this education was strongly influenced by his living together with cousins who were to him much more than cousins for all his life. After a period of intense reading, arousing a first interest in astronomy and physics, his passion for mathematics emerged and dominated his further life.

In 1938, he communicated his first results to Prof. J. Hadamard of the Academy of Sciences in Paris, who published one of his papers in the Academy's "Comptes Rendus" and encouraged him to proceed in his selected course of investigation. From this moment, Ferran Sunyer i Balaguer maintained a constant interchange with the French analytical school, in particular with Mandelbrojt and his students. In the following years, his results were published regularly. The limited space here does not, unfortunately, allow for a critical analysis of his scientific achievements. In the mathematical community his work, for which he attained international recognition, is well known.

Ferran's physical handicap did not allow him to write down any of his papers by himself. He dictated them to his mother until her death in 1955, and when, after a period of grief and desperation, he resumed research with new vigor, his cousins took care of the writing. His working power, paired with exceptional talents, produced a number of results which were eventually recognized for their high scientific value and for which he was awarded various prizes. These honours not withstanding, it was difficult for him to reach the social and professional position corresponding to his scientific achievements. At times, his economic situation was not the most comfortable either. It wasn't until the 9th of December 1967, 18 days prior his death, that his confirmation as a scientific member was made public by the División de Ciencias, Médicas y de Naturaleza of the Council. Furthermore, he was elected only as "de entrada", in contrast to class membership.

Due to his physical constraints, the academic degrees for his official studies were granted rather belatedly. By the time he was given the Bachelor degree, he had already been honoured by several universities! In 1960 he finished his Master's degree and was awarded the doctorate after the requisite period of two years as a student. Although he had been a part-time employee of the Mathematical Seminar since 1948, he was not allowed to become a full member of the scientific staff until 1962. This despite his actually heading the department rather than just being a staff member.

His own papers regularly appeared in the journals of the Barcelona Seminar, *Collectanea Mathematica*, to which he was also an eminent reviewer and advisor. On several occasions, he was consulted by the Proceedings of the American Society of Mathematics as an advisor. He always participated in and supported guest lectures in Barcelona, many of them having been prepared or promoted by him. On the occasion of a conference in 1966, H. Mascart of Toulouse publicly pronounced his feeling of being honoured by the presence of F. Sunyer i Balaguer, "the first, by far, of Spanish mathematicians".

At all times, Sunyer i Balaguer felt a strong attachment to the scientific activities of his country and modestly accepted the limitations resulting from his attitude, resisting several calls from abroad, in particular from France and some institutions in the USA. In 1963 he was contracted by the US Navy, and in the following years he earned much respect for the results of his investigations. "His value to the prestige of the Spanish scientific community was outstanding and his work in mathematics of a steady excellence that makes his loss difficult to accept" (letter of condolence from T.B. Owen, Rear Admiral of the US Navy).

Twice, Sunyer i Balaguer was approached by young foreign students who wanted to write their thesis under his supervision, but he had to decline because he was unable to raise the necessary scholarship money. Many times he reviewed doctoral theses for Indian universities, on one occasion as the president of a distinguished international board. The circumstances under which Sunyer attained his scientific achievements also testify to his remarkable human qualities. Indeed, his manner was friendly and his way of conversation reflected his gift for friendship as well as enjoyment of life and work which went far beyond a mere acceptance of the situation into which he had been born. His opinions were as firm as they were cautious, and at the same time he had a deep respect for the opinion and work of others. Though modest by nature, he achieved due credit for his work, but his petitions were free of any trace of exaggeration or undue self-importance. The most surprising of his qualities was, above all, his absolute lack of preoccupation with his physical condition, which can largely be ascribed to the sensible education given by his mother and can be seen as an indicator of the integration of the disabled into our society.

On December 27, 1967, still fully active, Ferran Sunyer i Balaguer unexpectedly passed away. The memory of his remarkable personality is a constant source of stimulation for our own efforts.

Translated from Juan Augé: Ferran Sunyer Balaguer. *Gazeta Matematica*, 1.a Serie – Tomo XX – Nums. 3 y 4, 1968, where a complete bibliography can be found.

La panorámica desde aquí arriba es impresionante. La cara N. de las Buitreras cae a pico sobre el cauce del río Dúrcal y se contempla la vertiente S., escasamente conocida, de los Alayos de Dilar...

(Jorge Garzón Gutiérrez, *Andar por Sierra Nevada*)

To My Father

Preface

This book is intended to present joint research with my colleagues Josep-Maria Peris, Jean-Pierre Ramis and Carles Simó. It is strongly based on our original papers but some effort has been made to follow a logical order rather then a historical one. Although the reader should be familiar with the methods of Hamiltonian systems and differential algebra, I have included two review chapters on most of the required material.

I learned Hamiltonian systems with Carles Simó and then differential Galois theory with Jean-Pierre Ramis. They encouraged me to write this monograph, read the manuscript and suggested many ways to improve it. I wish to express my deep gratitude to both of them.

I would like to thank Josep-Maria Peris for his help and correction of delicate parts of the manuscript.

I am very much indebted to the Ferran Sunyer i Balaguer Foundation and the Institut d'Estudis Catalans for their support to publish this monograph.

Juan J. Morales Ruiz

Contents

Chapter 1

Introduction

During recent years the search for non-integrability criteria for Hamiltonian systems based upon the behaviour of solutions in the complex domain has acquired more and more relevance.

We comment on some historical antecedents of our approach to the non-integrability of Hamiltonian systems.

We consider a *real* symplectic manifold M of dimension $2n$ and a Hamiltonian system X_H defined over it. Let Γ be a particular integral curve $z = z(t)$ (which is not an equilibrium point) of the vector field X_H. Then we can write the variational equation (VE) along Γ,

$$\dot{\eta} = \frac{\partial X_H}{\partial x}(z(t))\eta.$$

Using the linear first integral $dH(z(t))$ of the VE it is possible to reduce this variational equation (i.e., to rule out one degree of freedom) and to obtain the so-called normal variational equation (NVE) which, in suitable coordinates, can be written as

$$\dot{\xi} = JS(t)\xi,$$

where, as usual,

$$J = \begin{pmatrix} 0 & I \\ -I & 0 \end{pmatrix}$$

is the standard symplectic form of dimension $2(n-1)$.

Poincaré gave a non-integrability criterion based on the monodromy matrix of the VE along a periodic real integral curve: if there are k first integrals of the Hamiltonian system, independent over the integral curve, then k characteristic exponents must be zero. Moroever, if these first integrals are in involution, then $2k$ characteristic exponents must necessarily be zero ([87], pp. 192–198).

J. J. Morales Ruiz, *Differential Galois Theory and Non-Integrability*
of Hamiltonian, Systems, Modern Birkhäuser Classics,
DOI: 10.1007/978-3-0348-0723-4_1, © Springer Basel 1999

Furthermore, in Poincaré's work we can also find the relation between linear first integrals of the variational equation and solutions of this differential equation ([87], p. 168). In fact, Poincaré's results are intimately related to the reduction process from the VE to the NVE.

In 1888 S. Kowalevski obtained a new case of integrability of the rigid body system with a fixed point, imposing that the general solution is a meromorphic function of *complex* time. In fact, as part of her method, she proved that, except for some particular solutions, the only cases in which the general solution is a meromorphic function of time are Euler's, Lagrange's and Kowalevski's cases [57]. Lyapounov generalized the Kowalevski result and proved that, except for some particular solutions, the general solution is single-valued only in the above-mentioned three cases. His method is based on analysis of the variational equation along some known solutions [66, 63].

In 1963 Arnold and Krylov analyzed sufficient conditions for the existence of a single-valued (but not complex analytical!) first integral of a complex linear differential equation; and, under certain conditions, they proved the uniform distribution of values of the monodromy group on the corresponding invariant. Their proof is based upon the properties of the closure of the monodromy group considered to be contained in a linear Lie group [5]. We remark that this is not so far from the fact that the Galois group of a Fuchsian linear differential equation is the Zariski closure of the monodromy group (see Chapter 2 below).

In 1982 Ziglin [114] proved a non-integrability theorem for complex analytical Hamiltonian systems. He used the constraints imposed by the existence of some first integrals on the monodromy group of the normal variational equation along some complex integral curve. This result concerns branching of solutions: the monodromy group expresses the ramifications of solutions of the normal variational equation in the complex domain.

Let M be a *complex* analytic symplectic manifold of complex dimension $2n$ with a holomorphic Hamiltonian system X_H defined over it. Let Γ be the Riemann surface corresponding to an integral curve $z = z(t)$ (which is not an equilibrium point) of the vector field X_H. Then, as above, we can write the VE along Γ and the NVE of dimension $2(n-1)$,

$$\dot{\xi} = JS(t)\xi.$$

In general, if there are k analytical first integrals, including the Hamiltonian, independent over Γ and in involution, then, in a similar way, we can reduce the number of degrees of freedom of the VE by k. The resulting equation, which admits $n - k$ degrees of freedom, is also called the NVE [8]. Then we have the following result by Ziglin:

Theorem 1.1 ([114]) *Suppose that a Hamiltonian system admits $n - k$ additional analytical first integrals, independent over a neighborhood of Γ but not* *necessarily on* Γ *itself. We assume, moreover, that the monodromy group of the NVE contains a non-resonant transformation* g. *Then, any other element of the monodromy group of the NVE sends eigendirections of* g *into eigendirections of* g.

We say that a linear transformation $g \in Sp(m, \mathbf{C})$ (the monodromy group is contained in the symplectic group) is resonant if there exist integers $r_1, \ldots, r_m$ such that $\lambda_1^{r_1} \cdots \lambda^{r_m} = 1$ (λ_i are the eigenvalues of g).

Ziglin himself, in a second paper, applied his theorem to the rigid body and showed that, except for the three above-mentioned cases, this system is not completely integrable. He also studied the problem of the existence of an additional partial first integral and eventually included the Goryachev-Chaplygin case. Finally he applied his method to the Hénon-Heiles system and to a particular Yang-Mills field. For this last system, Ziglin proved the non-existence of a *local* meromorphic first integral independent of the Hamiltonian in any neighborhood of the hyperbolic equilibrium point [115]. In the present book we also obtain local non-integrability of some Hamiltonian systems in a neighborhood of an equilibrium point. But in our situation the equilibrium points can be degenerate (see below Section 4.3).

In 1985 Ito applied the Ziglin theorem to the non-integrability of a generalization of the Hénon-Heiles system [48]. From this date until today many papers have appeared on this subject. We shall comment briefly on some of them.

Yoshida published a series of papers about the application of Ziglin's theorem to some homogeneous two degree of freedom potentials with invariant planes. For such potentials he can project the normal variational equation over the Riemann sphere and obtain a *hypergeometric equation* [110, 111]. Later Churchill and Rod interpreted Yoshida's results geometrically as a reduction of the associated holomorphic connection by discrete symmetries ([24], see also [9, 25]). Several other papers are also oriented towards applications [44, 112, 47, 101, 23]. In Section 5.1, we will improve Yosida's results.

The differential Galois approach to Ziglin's theory appeared for the first time, independently, in [25] and [75, 81]. The papers [8], [26] and [82] followed. Two applications of the theory (developed in [25]) to non-academic examples are [58, 59]. A common limitation of these works is the restriction to Fuchsian variational equations (their singularities must be regular singular). Here we overcome this difficulty. Our basic idea is very simple: rather than working with the monodromy group, we work directly with the differential Galois group. Another problem inherent in Ziglin's original approach is the distinction between two types of first integrals: those that are useful for reduction and those are not. Of course, if we assume the involutivity and independence of all the integrals, then, from a theoretical point of view, this distinction is no longer relevant. In fact, if some integrals are independent over Γ, then the differential Galois theory itself allows an explicit process of reduction, in some sense.

In a way, up to now all known non-integrability criteria have not taken into account the involutivity hypothesis: only the independence of the first integrals is used. So, with exception of Poincaré's result mentioned above, we allow for the first time an obstruction to complete integrability in the Liouville sense. This means taking into acount not only the *number* of *independent* first integrals, as in the works of Ziglin and his followers, but also the fact that they are in *involution*.

We emphasize that this monograph must be considered as research on the connection between two different concepts of integrability: integrability of a Hamiltonian system X_H and integrability of the variational equation along a particular solution Γ of X_H. It is reasonable to suppose that if the flow of X_H has a regular behaviour, then the linearized flow along Γ given by the variational equation must also be regular. So, we can express simply our guiding idea at the beginning of our work on this problem: if the initial Hamiltonian system X_H is completely integrable then the VE must also be integrable but differently, in the sense of the differential Galois theory, i.e., the corresponding Picard-Vessiot extension must be a Liouvillian extension, or equivalently the identity component of the corresponding differential Galois group must be a solvable algebraic group.

In fact, we ultimately obtained a more precise result: in the complete integrability case the identity component of the differential Galois group of the VE is necessarily *abelian*, i.e., the abelian structure of the Poisson algebra of first integrals of X_H is "projected" at the linear level.

This monograph is divided into three parts. In part I, consisting of Chapters 2 and 3, we present the basic tools of the two integrability theories mentioned above.

Chapter 2 is devoted to explaining the necessary concepts and results of differential Galois theory. In general, these results are well known, but some of them were obtained in joint research of the author with J.P. Ramis (Proposition 2.4 and Theorems 2.4, 2.5) [77] and with C. Simó (Propositions 2.2, 2.3 and 2.6) [81, 82]. The proof of Proposition 2.4 and of Theorems 2.4, 2.5 is given in Appendices A and B. In Appendix C (about connections with structure groups) we also follow the paper [77] of the author with J.P. Ramis. These appendices are technically more difficult and abstract than the main body of the rest of the monograph.

In Chapter 3 we explain several concepts of complete integrability of Hamiltonian systems. As in differential Galois theory, there are several concepts of integrability depending on the coefficient field considered, for a Hamiltonian system several different definitions arise according to the degree of regularity of the first integrals involved and whether we consider the real or complex situation. Some definitions, as in the case of algebraic completely integrable systems, are based on the dynamical behaviour of the system. In Section 3.4 we give some properties of the Poisson algebra of rational functions. The results of this section are obtained as part of joint work of the author with J.P. Ramis and are the purely algebraic side of the main results of Chapter 4 [77].

Part II consists of Chapter 4 which is the central chapter of this monograph. There we explain the above mentioned result on the abelian character of the identity component of the Galois group. The results of this chapter were also obtained in a joint work of the author with J.P. Ramis [77].

Part III is devoted to applications. In Chapter 5 we apply our non-integrability result of Chapter 4 to three non-academic classical situations: homogeneous potentials, a cosmological model and a three body problem, obtaining not only new simple proofs of known results but many new results. This chapter is a reformulation of some joint results with J.P. Ramis [78], and we give an additional example of homogeneous potentials.

In Chapter 6 we analyze the non-integrability of a family of two degrees of freedom potentials with an invariant plane and a normal variational equation of Lamé type. Particular cases are the Hénon-Heiles family and the Toda family of three particles with two unequal masses. This chapter is part of joint work of the author with C. Simó [82].

Chapter 7 is devoted to a connection between the Galoisian non-integrability criterion of Chapter 4 and Lerman's real dynamical criterion of non-integrability, in a particular situation. This connection was conjectured by the author when he met Lerman some years ago at a meeting in Torun, Poland. The results of this chapter were obtained in a joint work with J.M. Peris [76].

In Chapter 8 we give some complementary applications and we formulate some conjectures that open new lines of research.

The reader may check that applications can be done using a unified and systematic approach:

1. Select a particular solution.
2. Write the VE and the NVE.
3. Check if the identity component of the differential Galois group of the NVE is abelian.

As we will see, step 2 is easy. In Chapter 4 we give an algorithm for obtaining the NVE from the VE. Step 3 is quite problematic in general, but fortunately, in some particular cases common to many applications, there exist efficient algebraic algorithms that can decide. The prototype is Kovacic's algorithm for second order equations. In all the applications that the author knows, step 1 (which is shared by all the classical proofs of non-integrability) is achieved by

the existence of a completely integrable subsystem, typically, by the existence
of an invariant plane. Of course this is in some sense unsatisfying from a philo-
sophical point of view: if a system is "as far as possible" from integrability,
then each integral curve will be pathological and our method does not work.
Anyway in such a case the classical methods fail for the same reason.

Chapter 2

Differential Galois Theory

The differential Galois theory for linear differential equations is the Picard-Vessiot Theory. In this theory there is a very nice concept of "integrability" i.e., solutions in closed form: an equation is integrable if the general solution is obtained by a combination of algebraic functions (over the coefficient field), exponentiation of quadratures and quadratures. Furthermore, all information about the integrability of the equation is coded in the identity component of the Galois group: the equation is integrable if, and only if, the identity component of its Galois group is solvable. It is a powerful theory in the sense that, in some favorable cases (for instance, for equations of order 2), it is possible to construct algorithms to determine whether a given linear differential equation is integrable or not.

We shall present only the essential definitions. Results shall be stated without proofs, unless the author has some contribution to them or if they are not easily found in the references. Three different approachs shall be used: ([12, 21, 50, 51, 54, 71, 69, 94, 102]): the classical approach, the Tannakian approach and the monodromy and Stokes's multipliers approach. As will become clear, all of them will be useful in this monograph.

In Sections 2.3, 2.4, 2.5 and 2.6 we will follow [77].

2.1 Algebraic groups

In this section the necessary results of linear algebraic groups are presented. An introduction to linear algebraic groups is given in [19]. For more information see the monographs [45, 14].

A linear algebraic group G (over $\mathbf{C}$) is a subgroup of $GL(m, \mathbf{C})$ whose matrix coefficients satisfy polynomial equations over $\mathbf{C}$. It has structures of an algebraic variety (non-singular) as well as of a group, and these two structures

J. J. Morales Ruiz, *Differential Galois Theory and Non-Integrability of Hamiltonian*, Systems, Modern Birkhäuser Classics, DOI: 10.1007/978-3-0348-0723-4_2, © Springer Basel 1999

are compatible: the group operation and taking of inverses are morphisms of algebraic varieties. We note that in a linear algebraic group there are two different topologies: the Zariski topology, where the closed sets are the algebraic sets, and the usual real topology. In particular, an algebraic group is a complex analytical Lie group and we can consider the Lie algebra of this group. Therefore the dimension of G is the dimension of the Lie algebra of G. Given a linear algebraic group G, the identity component (or the neutre component) G^0 is the (unique) irreducible component which contains the identity element of G.

We remark that an algebraic linear (or affine) group G is usually defined as an affine algebraic variety with a group structure, with the compatibility condition above: the group multiplication and taking of inverses are morphisms of algebraic varieties. Then, given a such G, there is a rational faithful representation of G as a closed subgroup of $GL(m, \mathbf{C})$, for some m, and we obtain the equivalence with our definition.

It is clear that the classical linear complex groups are linear algebraic groups. For instance $SL(n, \mathbf{C})$, $SO(n, \mathbf{C})$ (rotation group) and $Sp(n, \mathbf{C}) \subset Gl(2n, \mathbf{C})$ (symplectic group) are linear algebraic groups since they are defined by polynomial identities.

Proposition 2.1 *The identity component G^0 of a linear algebraic group G is a closed (with respect to the two above topologies) normal subgroup of G of finite index and it is connected with respect to the two above topologies. Furthermore the classes of G/G^0 are the irreducible connected components of G.*

We note that by the above proposition G^0 is also a linear algebraic group and the Lie algebra of G, $\mathrm{Lie}(G) = \mathcal{G}$ coincides with the Lie algebra of G^0, $\mathrm{Lie}(G^0) = \mathcal{G}$. As for every Lie group, G^0 is solvable (abelian) if, and only if, $\mathcal{G}$ is solvable (respectively abelian). Furthermore, G is connected if, and only if $G = G^0$.

The characterization of the connected solvable linear algebraic groups is given by the Lie-Kolchin theorem.

Theorem 2.1 (Lie-Kolchin Theorem) *A connected linear algebraic group is solvable if, and only if, it is conjugated to a triangular group.*

In the context of linear algebraic groups a torus is a group isomorphic to the multiplicative group $(\mathbf{C}^*)^k$. The dimension of the above torus is k. Equivalently, it is a linear algebraic group conjugated to a diagonal group. It is clear that a torus is connected and abelian.

Let G be a linear algebraic group. A maximal torus in G is a torus of maximal dimension contained in G. As a maximal torus is connected, it is contained in the identity component G^0.

Example. Let $Sp(n, \mathbf{C}) \subset Gl(2n, \mathbf{C})$ be the symplectic group. It is easy to see that the maximal tori in $Sp(n, \mathbf{C})$ are all the groups conjugated to

$$T = \{\mathrm{diag}(\lambda_1, \lambda_2, \ldots, \lambda_n, \lambda_1^{-1}, \lambda_2^{-1}, \ldots, \lambda_n^{-1}), \lambda_i \in \mathbf{C}^*, i = 1, 2, \ldots, n\}.$$

Indeed as is well known, the eigenvalues of the symplectic matrices $\sigma \in Sp(n, \mathbf{C})$ appear in pairs (λ, λ^{-1}) (see for instance [3]) and we get the above.

Given a subset $S \subset GL(n, \mathbf{C})$, let M be the group generated by S and G be the Zariski closure of the group M. By definition the group G is a linear algebraic group and we will say that this group is topologically generated by the set S. Sometimes we will emphasize in the difference between M and G and we will say that M is algebraically generated by S.

Since the examples of irreducible equations that we shall meet will be of second order and symplectic, we end this section with a classification of the algebraic subgroups of $SL(2, \mathbf{C})$. We shall need two lemmas.

Lemma 2.1 ([50]) *Let G be an algebraic group contained in $SL(2, \mathbf{C})$. Assume that the identity component G^0 of G is solvable. Then G is conjugate to one of the following types:*

(1) *G is finite,*

(2) $G = \left\{ \begin{pmatrix} \lambda & 0 \\ 0 & \lambda^{-1} \end{pmatrix}, \begin{pmatrix} 0 & -\beta^{-1} \\ \beta & 0 \end{pmatrix} \lambda, \beta \in \mathbf{C}^* \right\}$,

(3) *G is triangular.*

Lemma 2.2 *Let G be an algebraic subgroup of $SL(2, \mathbf{C})$ such that the identity component G^0 is not solvable. Then $G = SL(2, \mathbf{C})$.*

The last lemma is well known and it follows easily from consideration of the Lie algebra of $G \subset SL(2, \mathbf{C})$. Indeed, if G^0 is not solvable then the dimension of G must be equal to 3, because all 2-dimensional Lie algebras are solvable.

Proposition 2.2 ([81]) *Any algebraic subgroup G of $SL(2, \mathbf{C})$ is conjugated to one of the following types:*

1. *Finite, $G^0 = \{1\}$, where $1 = \begin{pmatrix} 1 & 0 \\ 0 & 1 \end{pmatrix}$.*

2. $G = G^0 = \left\{ \begin{pmatrix} 1 & 0 \\ \mu & 1 \end{pmatrix}, \mu \in \mathbf{C} \right\}$.

3. $G_k = \left\{ \begin{pmatrix} \lambda & 0 \\ \mu & \lambda^{-1} \end{pmatrix}, \lambda \text{ is a } k\text{-root of unity}, \mu \in \mathbf{C} \right\}$,

$$G^0 = \left\{ \begin{pmatrix} 1 & 0 \\ \mu & 1 \end{pmatrix}, \mu \in \mathbf{C} \right\}.$$

4. $G = G^0 = \left\{ \begin{pmatrix} \lambda & 0 \\ 0 & \lambda^{-1} \end{pmatrix}, \lambda \in \mathbf{C}^* \right\}$.

5. $G = \left\{ \begin{pmatrix} \lambda & 0 \\ 0 & \lambda^{-1} \end{pmatrix}, \begin{pmatrix} 0 & -\beta^{-1} \\ \beta & 0 \end{pmatrix} \lambda, \beta \in \mathbf{C}^* \right\}$,

$$G^0 = \left\{ \begin{pmatrix} \lambda & 0 \\ 0 & \lambda^{-1} \end{pmatrix}, \lambda \in \mathbf{C}^* \right\}.$$

6. $G = G^0 = \left\{ \begin{pmatrix} \lambda & 0 \\ \mu & \lambda^{-1} \end{pmatrix}, \lambda \in \mathbf{C}^*, \mu \in \mathbf{C} \right\}$.

7. $G = G^0 = SL(2, \mathbf{C})$.

Proof. Assume G to be infinite and conjugated to a triangular group, i.e., it is contained in the total triangular group (isomorphic to the semidirect product of the additive group $\mathbf{C}$ and of the multiplicative group $\mathbf{C}^*$)

$$\left\{ \begin{pmatrix} \lambda & 0 \\ \mu & \lambda^{-1} \end{pmatrix}, \lambda \in \mathbf{C}^*, \mu \in \mathbf{C} \right\}.$$

Let ψ be the morphism of algebraic groups

$$\psi : G \longrightarrow \mathbf{C}^*,$$

defined by

$$\psi \begin{pmatrix} \lambda & 0 \\ \mu & \lambda^{-1} \end{pmatrix} = \lambda.$$

If $\ker \psi$ is trivial then G must be the diagonal group

$$G = \left\{ \begin{pmatrix} \lambda & 0 \\ 0 & \lambda^{-1} \end{pmatrix}, \lambda \in \mathbf{C}^* \right\},$$

because then $G \approx \psi(G)$, $\psi(G)$ being an algebraic subgroup of the multiplicative group $\mathbf{C}^*$. But then $\psi(G)$ must be equal to $\mathbf{C}^*$ (the only possible non-trivial subgroups of $\mathbf{C}^*$ are the cyclic finite groups).

If $\ker \psi$ is non-trivial then, as it is (isomorphic to) an algebraic subgroup of the additive group $\mathbf{C}$, it is the total unipotent group

$$\left\{ \begin{pmatrix} 1 & 0 \\ \mu & 1 \end{pmatrix}, \mu \in \mathbf{C} \right\}.$$

Now as above we have two possibilities: either $\psi(G)$ is equal to the multiplicative group $\mathbf{C}^*$ or it is a finite cyclic group. The proposition follows from the two lemmas above. $\square$

The above proposition is analogous (but more precise: we need to know when the identity component of the Galois group is not only solvable, but abelian) to the proposition in [56], p. 7. We remark that the identity component G^0 is abelian in cases (1)–(5) and is solvable in cases (1)–(6).

2.2 Classical approach

A differential field K is a field with a derivative (or derivation) $\delta = '$, i.e., an additive mapping that satisfies the Leibniz rule. Examples are $\mathcal{M}(\overline{\Gamma})$ (meromorphic functions over a connected Riemann surface $\overline{\Gamma}$, the reason for this notation will be clear below: $\overline{\Gamma} - \Gamma$ will be the set of singular points of the linear differential equation) with a non-trivial meromorphic tangent vector field X as derivation, in particular $\mathbf{C}(z) = \mathcal{M}(\mathbf{P}^1)$ with $\frac{d}{dz}$ or $z\frac{d}{dz}$ as derivation, $\mathbf{C}\{x\}[x^{-1}]$ (convergent Laurent series), or $\mathbf{C}[[x]][x^{-1}]$ (formal Laurent series) with $x\frac{d}{dx}$ as derivation. We observe that there are some inclusions between the above differential fields.

We can define (differential) subfields, (differential) extensions in a direct way by requiring that inclusions must commute with the derivations. Analogously, a (differential) automorphism in K is an automorphism that commutes with the derivative. The field of constants of K is the kernel of the derivative. In the above examples $\mathbf{C}$ is such a kernel. From now on we will suppose that this is the case.

Let

$$\xi' = A\xi, \quad A \in Mat(m, K). \tag{2.1}$$

We shall proceed to associate to (2.1) the so-called Picard-Vessiot extension of K. The Picard-Vessiot extension L of (2.1) is an extension of K, such that if $u_1, \ldots, u_m$ is a "fundamental" system of solutions of the equation (2.1) (i.e., linearly independent over $\mathbf{C}$), then $L = K(u_{ij})$ (rational functions in K in the coefficients of the "fundamental" matrix $(u_1 \cdots u_m)$). This is the extension of K generated by K together with u_{ij}. We observe that L is a differential field (by (2.1)). The existence and unicity (except by isomorphism) of the Picard-Vessiot extensions is proved by Kolchin (in the analytical case, $K = \mathcal{M}(\overline{\Gamma})$, and this result is essentially the existence and uniqueness theorem for linear differential equations).

As in classical Galois theory, we define the Galois group of (2.1) $G := \mathrm{Gal}_K(L) = \mathrm{Gal}(L/K)$ as the group of all the (differential) automorphisms of L which leave fixed the elements of K. This group is isomorphic to an algebraic linear group over $\mathbf{C}$. We say that the extension L/K is normal if any element of L, invariant by the Galois group $\mathrm{Gal}_K(L)$, necessarily belongs to K. The Picard-Vessiot extensions are normal and by this property of the Picard-Vessiot extensions it is proved that the Galois correspondence (between groups and extensions) works well in this theory.

Theorem 2.2 *Let L/K be the Picard-Vessiot extension associated to a linear differential equation. Then there is a $1 - 1$ correspondence between the intermediary differential fields $K \subset M \subset L$ and the algebraic subgroups $H \subset G :=$*

$\mathrm{Gal}_K(L)$, *such that* $H = \mathrm{Gal}_M(L)$ *(the extension L/M is a Picard-Vessiot extension). Furthermore, we have*

(i) *The normal extensions M/K correspond to the normal subgroups $H \subset G$. Then the group G/H is a linear algebraic group, the extension M/K is a Picard-Vessiot extension and $G/H = \mathrm{Gal}_K(M)$.*

(ii) *Let F be a subgroup of G and K_F the subfield of L given by the elements of L fixed by F. Then $H := \mathrm{Gal}_{K_F}(L)$ is the Zariski closure (over the field of constants $\mathbf{C}$) of F (i.e., H is topologically generated by F).*

As a corollary, when we consider the algebraic closure $\overline{K}$ (of K in L), we obtain $\mathrm{Gal}_K(\overline{K}) = G/G^0$, where $G^0 = \mathrm{Gal}_{\overline{K}}(L)$ is the identity component of the Galois group G which corresponds to the transcendental part of the Picard-Vessiot extension, i.e., by definition, the extension $L/\overline{K}$ is the maximal transcendental extension between the extensions L/L_1, with L_1 an extension of K. If $\overline{K} = K$ (i.e., if $G = G^0$), we say that L/K is a purely transcendental extension.

Another consequence of Theorem 2.2 is that if $\Lambda \subset \overline{\Gamma}$ is a Riemann surface contained in $\overline{\Gamma}$ and L is a Picard-Vessiot extension of $\mathcal{M}(\overline{\Gamma})$, then $\mathrm{Gal}_{\mathcal{M}(\Lambda)}(L) \subset \mathrm{Gal}_{\mathcal{M}(\overline{\Gamma})}(L)$. We will apply this in Chapter 7. In a similar way, the local Galois group at a singular point $s \in \overline{\Gamma} - \Gamma$, $\mathrm{Gal}_{\mathbf{C}\{x\}[x^{-1}]}(L) := \mathrm{Gal}_{k_s}(L)$, is a subgroup of the global Galois group $\mathrm{Gal}_{\mathcal{M}(\Gamma)}(L)$ (as usual, we identify the germs of meromorphic functions at a singular point s with Laurent series centered at this point).

We will say that a linear differential equation is (Picard-Vessiot) integrable (or solvable) if we can obtain its Picard-Vessiot extension $K \subset L$ and, hence, the general solution, by adjunction to K of integrals, exponentiation of integrals or algebraic functions of elements of K. In other words, there exists a chain of differential extensions $K_1 := K \subset K_2 \subset \cdots \subset K_r := L$, where each extension is given by the adjunction of one element a, $K_i \subset K_{i+1} = K_i(a, a', a'', \ldots)$, such that a satisfies one of the following conditions:

(i) $a' \in K_i$,

(ii) $a' = ba$, $b \in K_i$,

(iii) a is algebraic over K_i.

The usual terminology is that the Picard-Vessiot extension is Liouvillian. Then, it can be proved that a linear differential equation is integrable if, and only if, the identity component of the Galois group, G^0, is a solvable group. In particular, if the identity component is abelian, the equation is integrable.

Furthermore, the relation between the monodromy and the Galois group is as follows.

Let $\overline{\Gamma} - \Gamma$ be the set of singular points of the equation i.e., the poles of the coefficients on $\overline{\Gamma}$. We recall that the monodromy group of the equation is a subgroup of the linear group, given by the image of a representation of the fundamental group $\pi_1(\Gamma)$ into the linear group $GL(m, \mathbf{C})$. This representation is obtained by analytical continuation of the solutions along the elements of $\pi_1(\Gamma)$ (see for instance [36]). The monodromy group is contained in the Galois group and if the equation is of Fuchsian class (i.e., it has regular singular singularities only), then the Galois group is dense in the monodromy group (Zariski topology) i.e., the Galois group is topologically generated by the monodromy group (see [69]). In the general case, Ramis found a generalization of the above and, for example, he proved that the Stokes matrices associated to an irregular singularity belong to the (local) Galois group (see Section 2.5 below). We will formulate a generalization of this result in Theorem 2.4 (see Appendix B for the proof).

A useful criterion for unimodularity is the following. The second order equation with coefficients p and q in a differential field K

$$\xi'' + p\xi' + q\xi = 0, \tag{2.2}$$

has a Galois group contained in $SL(2, \mathbf{C})$ if, and only if, $p = nd/d'$, for some $n \in \mathbf{Z}$, $d \in K$. To show this we note that for all σ in the Galois group, the Wronskian w belongs to K if, and only if, $w = \sigma(w) = \det(\sigma)w$, which is equivalent to $\det(\sigma) = 1$. We get this result by Abel's formula $w' + pw = 0$ (we take $w = Cd^n$, with $C \in \mathbf{C}$).

By the above criterion, the equation

$$\xi'' + g\xi = 0, \tag{2.3}$$

(where $g \in K$) has a Galois group contained in $SL(2, \mathbf{C})$. Now the classical change $v = -\xi'/\xi$ leads to the associated Riccati equation

$$v' = g + v^2. \tag{2.4}$$

Then

Proposition 2.3 ([81]) *If the equation $\xi'' + g\xi = 0$ is integrable then we are in one of the situations 1 to 6 of Proposition 2.2, and if we assume that the Galois group is not finite, then one has for the Riccati equation (2.2) the following:*
  1. *Cases 2, 3 and 6: it has exactly one solution in K.*
  2. *Case 4: it has two solutions in K.*
  3. *Case 5: it has two solutions in a quadratic extension of K but they do not belong to K.*

Proof. (A less detailed statement can be found in [50].) In cases 2, 3 and 6 there exists a solution, ξ_1, such that

$$\sigma\left(\frac{\xi'}{\xi}\right) = \frac{\xi'}{\xi},$$

for any σ in the Galois group. By the normality of the Picard-Vessiot extensions, one has that $v_1 = -\xi_1'/\xi_1$ belongs to K. Therefore v_1 is a solution of the Riccati equation in K. Let us assume that there is another solution, v_2 of the Riccati equation in K. Let ξ_2 be defined by $v_2 = -\xi_2'/\xi_2$. Then $\{\xi_1, \xi_2\}$ is a fundamental system of solutions of $\xi'' + g\xi = 0$, because

$$v_2 - v_1 = \frac{w}{\xi_1 \xi_2},$$

w being the Wronskian of $\{\xi_1, \xi_2\}$. But for each element

$$\sigma = \begin{pmatrix} \alpha & \gamma \\ \beta & \delta \end{pmatrix}$$

in the Galois group G, $\sigma(v_i) = v_i$, $i = 1, 2$, implies that G is diagonal. Indeed, if

$$\frac{\xi}{\xi'} = \sigma(\frac{\xi}{\xi'}) = \frac{\alpha \xi_1' + \beta \xi_2'}{\alpha \xi_1 + \beta \xi_2},$$

then $\beta w = 0$ and, therefore, $\beta = 0$. In an analogous way we obtain $\gamma = 0$. Then G would be diagonal and this contradicts the hypothesis.

In case 4, let $\{\xi_1, \xi_2\}$ be a fundamental system of solutions such that $\sigma(\xi_1) = \lambda \xi_1$, $\sigma(\xi_2) = \lambda^{-1}\xi_2$, with σ an element of the Galois group. Hence $\sigma(\xi'/\xi) = \xi'/\xi$, and by normality we get $v_i = -\xi'/\xi \in K$, for $i = 1, 2$. Of course one has $v_1 \neq v_2$ because

$$v_2 - v_1 = \frac{w}{\xi_1 \xi_2}.$$

In case 5, G/G^0 is the Galois group of the quadratic extension $\overline{K}/K$ and G^0 is the Galois group of the extension $L/\overline{K}$ (see the remark after Theorem 2.2). As G^0 is diagonal, the proof that the Riccati equation has two different solutions v_1, v_2 in $\overline{K}$ proceeds as in case 4. They do not belong to K because then G would be diagonal. We note that, in case 5, if the Riccati equation has one solution $\overline{K}$ then it has two solutions in $\overline{K}$. Indeed, let $v_1 = \kappa + \sqrt{\omega}$ be a solution in $\overline{K}$, with $\kappa, , \omega \in \overline{K}$, $\sqrt{\omega} \notin K$, then $v_2 = \kappa - \sqrt{\omega}$ is another solution in $\overline{K}$. □

We remark that although the above proposition is closely related to Kovacic's algorithm (see Section 2.6), we establish it in an independent way, because for some particular equations (as for Lamé's equation, see Section 2.8.4) it gives us good results without using all the machinery of Kovacic's algorithm.

We finish this section with a remark about differential extensions by quadratures. Let L/K be a differential extension by integrals, i.e., $L = K(a_1, a_2, \ldots, a_s)$, where $a_i' \in K$, $i = 1, 2, \ldots, s$. Then L/K is a Picard-Vessiot extension and the Galois group $\mathrm{Gal}(L/K)$ is isomorphic to an additive group

$G_a^r := (\mathbf{C}^r, +)$, for some $r \leq s$. For $s = 1$ (see [50, 71]), the corresponding linear differential equation is a second order equation. For arbitrary s, we write the corresponding linear differential equation as a direct sum of s second order equations and we obtain the linear representation of the Galois group as an additive subgroup of the unipotent linear group contained in $GL(2s, \mathbf{C})$. In particular, $\mathrm{Gal}_K(L)$ is connected, and L/K is a purely transcendental extension.

2.3   Meromorphic connections

Linear connections are the intrinsic version of systems of linear differential equations. Moreover, with connections it is possible to work with necessarily non-trivial fibre bundles. A good reference for this section is [104] (see also [29, 30, 51, 69]).

Let Γ be a (connected) Riemann surface. We denote by $\mathcal{O}_\Gamma$ its sheaf of holomorphic functions, by Ω_Γ its sheaf of holomorphic 1-forms (corresponding to the canonical bundle) and by $\mathcal{X}_\Gamma$ its sheaf of holomorphic vector fields. We will identify vector fields with derivations on $\mathcal{O}_\Gamma$. We have a sheaf structure of Lie-algebras on $\mathcal{X}_\Gamma$. There exist, clearly, natural structures of $\mathcal{O}_\Gamma$-modules on Ω_Γ and $\mathcal{X}_\Gamma$, respectively. There exists a natural map (contraction)

$$\Omega_\Gamma \otimes_{\mathcal{O}_\Gamma} \mathcal{X}_\Gamma \to \mathcal{O}_\Gamma,$$

$$\omega \otimes v \to \langle \omega, v \rangle.$$

Let V be a holomorphic vector bundle of rank m on Γ. We denote by $\mathcal{O}_V$ its sheaf of holomorphic sections. Then a *holomorphic* connection is by definition a map

$$\nabla : \mathcal{O}_V \to \Omega_\Gamma \otimes_{\mathcal{O}_\Gamma} \mathcal{O}_V,$$

satisfying the Leibniz rule

$$\nabla(v + w) = \nabla v + \nabla w$$

$$\nabla f v = df \otimes v + f \nabla v,$$

where v, w are holomorphic sections of the fibre bundle V and f is a holomorphic function.

By definition a section v of the fibre bundle V is horizontal for the connection ∇ if $\nabla v = 0$.

If the connection ∇ is fixed, then to each holomorphic vector field X over Γ, we can associate the covariant derivative along X

$$\nabla_X : \mathcal{O}_V \to \mathcal{O}_V,$$
$$\nabla_X : v \to \langle \nabla v, X \rangle.$$

It is clearly a **C**-linear map. If we denote by $\mathrm{End}_{\mathbf{C}}(\mathcal{O}_V)$ the sheaf of spaces of **C**-linear endomorphisms of the sheaf of complex vector spaces $\mathcal{O}_V$, then we get a map

$$\nabla : \mathcal{X}_\Gamma \to \mathrm{End}_{\mathbf{C}}(\mathcal{O}_V),$$

$$X \mapsto \nabla_X,$$

such that

$$\nabla_X(v + w) = \nabla_X v + \nabla_X w,$$

$$\nabla_X(fv) = X(f)v + f\nabla_X v, \quad f \in \mathcal{O}_\Gamma.$$

We are going to compute ∇ in local coordinates. Let X be a holomorphic vector field over an open subset U of the Riemann surface Γ. Restricting U, if necessary, we can suppose that there exists a holomorphic local coordinate t over U such that

$$X = \frac{d}{dt}.$$

Let $e = \{e_1, \ldots, e_m\}$ be a holomorphic frame of U, i.e., the data of m holomorphic sections of V over U, such that $e_1(p), \ldots, e_m(p) \in V_p$ are linearly independent at every point $p \in U$. Then we can set

$$\nabla e_j = -\sum_{i=1}^{m} a_{ij} e_i,$$

(a_{ij}) being a square matrix of order m whose entries are holomorphic functions over U. We write $\nabla e = -Ae$.

The matrix $A = (a_{ij})$ is by definition the connection matrix and it determines completely the connection: if v is a holomorphic section over U, then we can write it in coordinates

$$v = \sum_{i=1}^{m} \xi_i e_i,$$

where the ξ_i's are holomorphic functions over U, and we have

$$\nabla v = \sum_{i=1}^{m} \left(\frac{d\xi_i}{dt} - \sum_{j=1}^{m} a_{ij} \xi_j \right) e_i,$$

i.e., the connection ∇ is represented in the local coordinate t and the frame e by the linear differential operator

$$\nabla := \nabla_{\frac{d}{dt}} = \frac{d}{dt} - A.$$

Hence, we can associate to the solutions $\xi \in \mathcal{O}_U^m$ of the linear differential system

$$\frac{d\xi_i}{dt} = \sum_{j=1}^{m} a_{ij}\xi_j, \quad i = 1, \ldots, m,$$

the horizontal sections v of the connection

$$\nabla v = 0.$$

More precisely the map

$$\xi \mapsto \sum_{i=1}^{m} \xi_i e_i$$

induces an isomorphism of m-dimensional complex vector spaces between the space of solutions and the space of horizontal sections.

In fact we are interested not only in differential equations (or systems) with *holomorphic* coefficients, but also in differential equations (or systems) with *meromorphic* coefficients. Therefore we need to extend the above concept of holomorphic connection in order to deal with poles and consequently to introduce *meromorphic* connections. We shall follow Section 4 of [104] (a more elaborated analysis in the context of free coherent sheaves can be found in [67]).

Let $\overline{\Gamma}$ be a Riemann surface and V a *holomorphic* vector bundle on $\overline{\Gamma}$. In our applications, the following specific conditions will hold. Let $\Gamma \subset \overline{\Gamma}$ be an open subset such that $S = \overline{\Gamma} - \Gamma$ is a discrete subset (the singular set). We will consider meromorphic sections of the bundle V, and in general we will limit ourselves to sections whose restriction to Γ is *holomorphic*. Then at any point $s \in S$ their components, in coordinates with respect to a holomorphic local frame, are meromorphic functions in a neighborhood U_s, which are holomorphic on $U_s - \{s\}$, with a pole at s. Using a local holomorphic coordinate t, vanishing at s, we can identify these functions with elements of the field $\mathbf{C}\{t\}[t^{-1}]$. That is the field $\mathbf{C}\{t\}[t^{-1}]$ with the field k_s of germs at s of meromorphic functions.

We denote by $\mathcal{M}_{\overline{\Gamma}}$ the sheaf of meromorphic functions over $\overline{\Gamma}$, by $\mathcal{M}_{\overline{\Gamma}}^1 = \mathcal{M}_{\overline{\Gamma}} \otimes_{\mathcal{O}_{\overline{\Gamma}}} \Omega_{\overline{\Gamma}}$ the sheaf of meromorphic 1-forms, and by $\mathcal{L}_{\overline{\Gamma}} = \mathcal{M}_{\overline{\Gamma}} \otimes_{\mathcal{O}_{\overline{\Gamma}}} \mathcal{X}_{\overline{\Gamma}}$ its sheaf of meromorphic vector fields. We have a sheaf structure of Lie algebras on $\mathcal{L}_{\overline{\Gamma}}$. Clearly there exist natural structures of sheaves of $\mathcal{M}_{\overline{\Gamma}}$-vector spaces on $\mathcal{M}_{\overline{\Gamma}}^1$ and $\mathcal{L}_{\overline{\Gamma}}$, respectively. There exists a natural map (contraction)

$$\mathcal{M}_{\overline{\Gamma}}^1 \otimes_{\mathcal{M}_{\overline{\Gamma}}} \mathcal{L}_{\overline{\Gamma}} \to \mathcal{M}_{\overline{\Gamma}},$$

$$\mu \otimes v \to \langle \mu, v \rangle.$$

Let V be a holomorphic vector bundle of rank m on $\overline{\Gamma}$. Then a *meromorphic* connection on V is by definition a map

$$\nabla : \mathcal{M}_V \to \mathcal{M}^1_{\overline{\Gamma}} \otimes_{\mathcal{M}_{\overline{\Gamma}}} \mathcal{M}_V,$$

satisfying the Leibniz rule

$$\nabla(v + w) = \nabla v + \nabla w$$

$$\nabla fv = df \otimes v + f\nabla v,$$

where v, w are holomorphic sections of the fibre bundle V and f is a meromorphic function.

If the meromorphic connection ∇ is fixed, then to each meromorphic vector field X over Γ we can associate the covariant derivative along X

$$\nabla_X : \mathcal{M}_V \to \mathcal{M}_V,$$

$$\nabla_X : v \mapsto \langle \nabla v, X \rangle.$$

It is clearly a **C**-linear map. Then if we denote by $\mathrm{End}_{\mathbf{C}}(\mathcal{M}_V)$ the sheaf of **C**-linear endomorphisms of the sheaf of complex vector spaces $\mathcal{M}_V$, we get a map

$$\nabla : \mathcal{L}_{\overline{\Gamma}} \to \mathrm{End}_{\mathbf{C}}(\mathcal{M}_V),$$

$$X \mapsto \nabla_X,$$

such that

$$\nabla_X(v + w) = \nabla_X v + \nabla_X w,$$

$$\nabla_X(fv) = X(f)v + f\nabla_X v, \quad f \in \mathcal{M}_\Gamma.$$

Let ∇ be a meromorphic connection over $\overline{\Gamma}$. We will say that it is *holomorphic* at a point $p \in \overline{\Gamma}$ if, for every germ at p of the *holomorphic* vector field X, the space of germs at p of *holomorphic* sections of the fibre bundle V is invariant by the covariant derivative ∇_X. Later we will consider connections that are meromorphic on $\overline{\Gamma}$ and holomorphic on Γ. They can have poles on the singular set S.

If we want to compute in local coordinates in a neighborhood of a singular point $s \in S$, then we choose a holomorphic coordinate t at s (vanishing at s) and we write our given vector field $X = f(t)\frac{d}{dt}$, where $f \in k_s$ (in general we cannot write X as $\frac{d}{dt}$, because the field X may vanish or admit a pole at the point s, as we shall see later in the applications). Then using a holomorphic frame e of V as above, we get a differential system

$$\nabla = f(t)\frac{d}{dt} - A(t).$$

We can introduce the meromorphically equivalent differential system

$$\frac{d}{dt} - B(t),$$

where $B = f^{-1}A$ is a *meromorphic* matrix over U.

We denote the field of global meromorphic functions over $\overline{\Gamma}$ by $k_{\overline{\Gamma}}$. It is important to notice that every holomorphic fibre bundle over a Riemann surface $\overline{\Gamma}$ is *meromorphically* trivial over $\overline{\Gamma}$ (i.e., globally, see Appendix A). Therefore its space of global meromorphic sections is isomorphic to some $k_{\overline{\Gamma}}^{m}$. In particular, we can choose a non-trivial meromorphic vector field X over $\overline{\Gamma}$. It will define a *derivation* δ over the field $k_{\overline{\Gamma}}$ and we will get a differential field $(k_{\overline{\Gamma}}, \delta)$. If V is a holomorphic vector bundle over $\overline{\Gamma}$ and if $\mathcal{M}(\overline{\Gamma}) \approx k_{\overline{\Gamma}}^{m}$ is its $k_{\overline{\Gamma}}$-vector space of meromorphic sections, then the covariant derivative ∇_X induces a **C**-linear endomorphism of the space $\mathcal{M}(\overline{\Gamma})$ and therefore it can be interpreted as a **C**-linear endomorphism of the space $k_{\overline{\Gamma}}^{m}$. We can choose as a local coordinate t over $\overline{\Gamma}$ a non-trivial global meromorphic function over $\overline{\Gamma}$ (it will be a true local coordinate, i.e., a local biholomorphism, but perhaps over a discrete subset). We can write $X = f(t)\frac{d}{dt}$, where $f \in k_{\overline{\Gamma}}$. Then we can choose a global meromorphic frame of V over $\overline{\Gamma}$, that is a set $e = \{e_1, \ldots, e_m\}$ of meromorphic sections of V inducing a true holomorphic frame over a non-trivial open subset (necessarily dense). Finally, proceeding as above, we can interpret our connection as a global meromorphic differential system

$$\nabla = f(t)\frac{d}{dt} - A(t), \quad \text{or equivalently} \quad \frac{d}{dt} - B(t),$$

where $B = f^{-1}A$ is a global meromorphic matrix whose entries belong to $k_{\overline{\Gamma}}$.

In the preceding process it is in general necessary to introduce new poles. We will keep our notations, always denoting by S the new singular set and by Γ the new regular set (i.e., the set S can be bigger than the set of poles of our connection).

We will also need meromorphic connections on *meromorphic* bundles over a Riemann surface $\overline{\Gamma}'$. It is easy, using Appendix A, to adapt the preceding definitions. We leave the details to the reader. In our applications the more general situation will be the following. The symbol ∇ will be a meromorphic connection on a meromorphic bundle over $\overline{\Gamma}'$. By restriction, we will get a meromorphic connection on a holomorphic bundle over an open dense subset $\overline{\Gamma} \subset \overline{\Gamma}'$, and by a new restriction a holomorphic connection on a holomorphic bundle over an open dense subset $\Gamma \subset \overline{\Gamma}$. The sets $\overline{\Gamma} - \Gamma$ and $\overline{\Gamma}' - \overline{\Gamma}$ will be discrete (frequently finite in the applications) subsets and they will correspond to the introduction of *equilibrium points* and *points at infinity*, respectively.

In the rest of this section we fix the (connected) Riemann surface $\overline{\Gamma}$, and the non-trivial meromorphic vector field X over $\overline{\Gamma}$. We interpret this field as a derivation on the field of global meromorphic functions $k_{\overline{\Gamma}} = \mathcal{M}(\overline{\Gamma})$ over $\overline{\Gamma}$. As we explained above, we can consider a meromorphic vector bundle as a vector space over $k_{\overline{\Gamma}}$.

From a given meromorphic connection ∇ defined on the vector bundle V, we can obtain an infinite number of induced meromorphic connections ([29, 30, 51, 69, 104]) by natural geometric processes. The idea is to extend naturally the connection to the tensor levels by requiring that the Leibniz rule be satisfied by the tensor products ($\nabla(u \otimes v) = \nabla u \otimes v + u \otimes \nabla v$) and that the action on a direct sum is the evident one (i.e., $\nabla(U \oplus V) = \nabla U \oplus \nabla V$). So, we can construct connections: ∇^*, $\otimes^k \nabla$, $\wedge^k \nabla$, $S^k \nabla$, acting on the bundles V^*, $\otimes^k V$, $\wedge^k V$, $S^k V$, respectively. By definition, $\otimes^0 V$ is the field of meromorphic functions and we endow it with the connection X (interpreted as a derivation on this field). With all these constructions we can build various direct sums and we can iterate the process. So, for example, $\wedge^3(\nabla^* \oplus S^2 \nabla)$ is an induced connection. If a subbundle is invariant by a connection, this connection is by definition a subconnection. We can also introduce subconnections and quotients in our machinery.

We observe the similarity of the above definitions to derivations in differential geometry (Lie derivative, etc...). This is not merely a coincidence as we will see in Section 4.1, where we will consider a connection as a Lie derivative.

In a natural way we can generalize the above in order to consider constructions using a family of given connections. For instance, let ∇_1 and ∇_2 be two meromorphic connections over the vector bundles V_1 and V_2, respectively. The tensor product $\nabla_1 \otimes \nabla_2$ is defined by the Leibniz rule as above, $\nabla_1 \otimes \nabla_2(u \otimes v) = \nabla_1 u \otimes v + v \otimes \nabla_2 v$, where $v \in V_1$ and $u \in V_2$. In an analogous way we define the direct sum of connections, etc.... Finally, we get the tensor category of the meromorphic connections over $\overline{\Gamma}$. The homomorphisms of this category are defined in the following way. A homomorphism ϕ between ∇_1 and ∇_2 is a homomorphism of the underlying vector spaces (over the field $k_{\overline{\Gamma}}$) $\phi : V_1 \to V_2$, such that $\phi \nabla_1 = \nabla_2 \phi$ (for more details and formal definitions, which are not needed here, the interested reader can look at [29]). Now it is clear how to extend the usual definitions on homomorphisms of vector spaces to homomorphisms of connections. For instance, an exact sequence of connections is given by an exact sequence of vector spaces, where the homomorphisms that define the sequence are homomorphisms of connections.

Now, we will obtain the connection matrices for some examples.

Example 1. The dual connection ∇^* is defined from the Leibniz rule by

$$X\langle \alpha, v \rangle = \langle \nabla^* \alpha, v \rangle + \langle \alpha, \nabla v \rangle,$$

where $v \in V$, $\alpha \in V^*$, and $\langle, \rangle$ denotes the duality. If e and e^* are dual frames in V and V^*, respectively, then we have

$$\langle \nabla^* e^*, e \rangle = \frac{d}{dt} \langle e^*, e \rangle + \langle e^*, eA \rangle = \langle e^* A^t, e \rangle,$$

A is the connection matrix of ∇ in the frame e, i.e., $\nabla e = -Ae$. Hence, we have obtained just the adjoint differential equation: the adjoint differential equation of

$$\frac{d\xi}{dt} = A\xi \quad \text{is by definition} \quad \frac{d\eta}{dt} = -A^t \eta.$$

We observe that, in order for $\alpha = \sum_{i=1}^{m} \eta_i e_i^*$ to be a linear first integral of

$$\nabla v = 0,$$

it is necessary and sufficient that

$$\nabla^* \alpha = 0.$$

This is a well-known property of the adjoint. In a similar way, it is possible to prove that the horizontal sections of $S^k \nabla^*$ are the homogeneous polynomial first integrals of the linear equation defined by the initial connection on V.

It is usual to write ∇ instead of ∇^*, $\otimes^k \nabla$, etc. . . , if the vector spaces on which they act are clear enough. We will follow this convention.

Example 2. The connection $\wedge^m \nabla$ $(\dim V = m)$ is defined by

$$\nabla(v_1 \wedge \cdots \wedge v_m) = \sum_{i=1}^{m} v_1 \wedge \cdots \wedge \nabla v_i \wedge \cdots \wedge v_m.$$

Then $\nabla(e_1 \wedge \cdots \wedge e_m) = \text{tr}A \; e_1 \wedge \cdots \wedge e_m$. We have obtained the differential equation for the determinant of a fundamental matrix, i.e., the so-called Jacobi-Abel formula ($v_1 \wedge \cdots \wedge v_m = \det(v_1, \ldots, v_m) e_1 \wedge \cdots \wedge e_m$).

Example 3. Let $0 \longrightarrow (V_1, \nabla_1) \longrightarrow (V, \nabla) \longrightarrow (V_2, \nabla_2) \longrightarrow 0$ be an exact sequence of connections. In other words the connection (V_1, ∇_1) is a subconnection of (V, ∇) (i.e., isomorphic to the restriction of ∇ over an invariant subspace of V by ∇), and (V_2, ∇_2) is isomorphic to the "normal" connection $(V/V_1, \tilde{\nabla})$ to (V_1, ∇_1), defined in the natural way. It is easy to verify that this normal connection is well defined. Then if we take a basis $e_1, \ldots, e_k, e_{k+1}, \ldots, e_n$ of V such that $e_1, \ldots, e_k$ is a basis of V_1, the matrix of the connection ∇ (we write in a more informal way ∇ instead of (V, ∇), etc. . .) is given by

$$\begin{pmatrix} A_1 & B \\ 0 & A_2 \end{pmatrix},$$

A_1 and A_2 being the matrices of the connections ∇_1 and ∇_2 respectively.

Now the method for solving the linear equation of (V, ∇) is the following. Let U_1, U_2 be fundamental matrices of the connections ∇_1, ∇_2 respectively, then a fundamental matrix of ∇ is given by

$$\begin{pmatrix} U_1 & V \\ 0 & U_2 \end{pmatrix}.$$

By writing explicitly the differential equation of U,

$$\frac{dU}{dt} = AU,$$

it is clear that the matrix V is obtained from U_1 and U_2 by the method of variation of constants. Then we have the chain of Picard-Vessiot extensions

$$K \subset K(U_1) \subset K(U_1, U_2) \subset K(U_1, U_2, V),$$

the last one being obtained by variation of constants.

As we will see later in Chapter 4, a similar method (but more involved due to the additional structure given by the symplectic form) will be used in order to reduce the variational equation (along a particular solution of a Hamiltonian system) to the normal variational equation. This will be also useful in Chapter 8.

In this book we are mainly interested in the following particular vector bundles and connections. A (meromorphic) symplectic vector bundle is a (meromorphic) vector bundle V such that there is a holomorphic section $\Omega \in \wedge^2 V^*$ whose restrictions to the fibres of V are not degenerated (the rank m of V is $2n$).

Then we have also the following result on the trivialization of a symplectic vector bundle.

Proposition 2.4 ([77]) *A symplectic vector bundle V over a Riemann surface is (symplectic) meromorphically trivial (i.e., there exists a global symplectic frame given by meromorphic sections).*

We will give a proof of the above proposition in Appendix A.

As above we denote by $k_{\overline{\Gamma}}$ the field of meromorphic functions over $\overline{\Gamma}$. We denote by $\mathcal{E}$ the $k_{\overline{\Gamma}}$-vector space of global meromorphic sections of V. The form Ω induces a $k_{\overline{\Gamma}}$-bilinear antisymmetric map

$$\Omega : \mathcal{E} \otimes \mathcal{E} \to k_{\overline{\Gamma}}, \qquad (v, w) \mapsto \Omega(v, w).$$

If v, w are holomorphic sections of V in a neighborhood of a point $p \in \overline{\Gamma}$, then $\Omega(v, w)(p) = \Omega(v(p), w(p)) \in \mathbf{C}$. Consequently the $k_{\overline{\Gamma}}$-bilinear map

$$\Omega; \mathcal{E} \otimes \mathcal{E} \to k_{\overline{\Gamma}}$$

is non-degenerate.

For many applications, we can identify the symplectic bundle V with the symplectic vector space $\mathcal{E}$ over the field $k_{\overline{F}}$. In this situation all the purely algebraic results on symplectic vector spaces over the numerical fields $\mathbf{R}$ or $\mathbf{C}$ remain also true [6]. In particular, there are symplectic bases i.e., canonical frames given by global meromorphic sections, and, with respect to a symplectic base, Ω is represented by the canonical form

$$J = \begin{pmatrix} 0 & I \\ -I & 0 \end{pmatrix}.$$

Furthermore, changes of symplectic bases are given by elements of the symplectic group $Sp(n, k_{\overline{F}}) \subset GL(2n, k_{\overline{F}})$.

By definition we will say that a (holomorphic or more generally meromorphic) connection ∇ over the *symplectic* bundle V (or (∇, V, Ω) in a more formal way) is *symplectic* if Ω is a horizontal section of $\wedge^2 \nabla^*$, i.e., it satisfies $\nabla \Omega = 0$ (for a related definition see [8]). Then, it is easy to see that, after a choice of coordinates, if we compute the connection matrix A of ∇ in a symplectic frame e, it satisfies

$$A^t J + J A = 0$$

(to show this, it is sufficient to note that $0 = \nabla \Omega = \nabla(e^* \otimes Je^{*t})$). This condition is equivalent to the existence of a meromorphic symmetric matrix S such that $A = JS$, and the matrix A belongs to the Lie algebra of the symplectic Lie group with coefficients in the field $k_{\overline{F}}$. Then the equation

$$\nabla v = 0$$

is the intrinsic expression of the linear Hamiltonian system

$$\dot{\xi} = JS\xi,$$

where $\xi = (\xi_1, \ldots, \xi_{2n})^t$ are the coordinates of v in the symplectic base and, as usual in dynamical systems, we denote the temporal derivative by a dot.

Conversely, if the matrix of the connection ∇ computed in a symplectic frame is symplectic, then $\nabla \Omega = 0$ and this connection is symplectic. Therefore our definition of a symplectic connection is equivalent to the definition of a connection with structure group $G = Sp(2n; \mathbf{C})$ given in Appendix A.

All the above constructions remain valid if we start with a local meromorphic connection on the vector space V over the field $\mathbf{C}\{t\}[t^{-1}]$ with the suitable dictionary: $\frac{d}{dt}$ instead X, etc.…

2.4   The Tannakian approach

We present now the Galois theory from the intrinsic connection perspective
[21, 29, 51, 69]. Let (V, ∇) be, as in the above section, a meromorphic connection over a fibre bundle of rank m. Then, we consider the horizontal sections,
$\mathrm{Sol}\,\nabla := \mathrm{Sol}_{p_0}\,\nabla$ of this connection at a fixed non-singular point $p_0 \in \Gamma$ (they
correspond to solutions of the corresponding linear equation). By the general
existence theory of linear differential equations, $\mathrm{Sol}\,\nabla$ is a vector space over $\mathbf{C}$
of dimension m (if we consider the solutions in a simply connected domain that
contain p_0). Then the mapping

$$(V, \nabla) \longrightarrow \mathrm{Sol}\,\nabla$$

is called a functor fibre (it is a functor between the tensor category of the
meromorphic connections and the tensor category of complex vector spaces).

Now, as in the previous section, we obtain the family of tensor constructions: $(V, \nabla), (V^*, \nabla^*)$, etc..., from a given connection. In this family we include the subconnections. A subconnection of a construction $(C(V), C(\nabla))$ is
an object $(W, C(\nabla)_{|W})$, W being a subbundle of $C(V)$ invariant by $C(\nabla)$. The
next step is to consider the corresponding spaces of solutions by the functor Sol,
for all the elements of this extended family. Then $C(\mathrm{Sol}\,\nabla) = \mathrm{Sol}(C\nabla)$, and the
Galois group of the initial connection (V, ∇), $\mathrm{Gal}\,\nabla$, is defined as the subgroup
of $GL(\mathrm{Sol}\,\nabla) \approx GL(m,\ C)$, which leaves invariant the spaces corresponding to
all constructions $C(V)$. We remark that $GL(\mathrm{Sol}\,\nabla)$ acts on any construction by
the usual pull-back. The key point is that the above group is isomorphic (as an
algebraic group) to the Galois group G of the corresponding linear equation.
This approach to the Picard-Vessiot theory is called the Tannakian point of
view.

Example. Let (V, ∇, Ω) be a symplectic connection with rank $V = 2n$ and
X the holomorphic vector field over $\overline{\Gamma}$. We make the construction $(\mathcal{M}_\Gamma(\overline{\Gamma}) \oplus
\wedge^2 V^*, X \oplus \wedge^2 \nabla^*)$, $\mathcal{M}_\Gamma(\overline{\Gamma})$ being the (global) meromorphic functions over $\overline{\Gamma}$,
holomorphic on Γ. The line subbundle generated by $1 + \Omega$, $\mathcal{M}_\Gamma(\overline{\Gamma})(1 + \Omega)$,
is invariant, because $\nabla\Omega = 0$ and $\nabla(f(1 + \Omega)) = X(f)(1 + \Omega)$, $f \in \mathcal{M}_\Gamma(\overline{\Gamma})$.
Hence, the corresponding construction by Sol, $\mathbf{C}(1 + \Omega_0)$ (Ω is a horizontal
section of $\wedge^2 \nabla^*$) is invariant by the Galois group. Therefore, the Galois group
is contained in the symplectic group $Sp(\mathrm{Sol}(V)) \approx Sp(n, \mathbf{C})$. A different proof
of this in a more general context will be given in Appendix C.

2.5 Stokes multipliers

The objective now is to state a theorem of Ramis which relates the Picard-Vessiot theory with the Stokes multipliers at an irregular singular point [90, 69, 74, 16]. For simplicity, we will explain only the main concepts necessary to understand the theorem, for the case of a second order differential equation (equivalently, for a system of dimension two). The reader can find a good introduction in [74] and the complete proof is in [16].

We start with the local case and we will consider that the singular point is at infinity, $x_0 = \infty$. Furthermore, we denote by $\hat{K} := \mathbf{C}[[x^{-1}]][x]$, $K := \mathbf{C}\{x^{-1}\}[x]$, the field of formal and convergent Laurent series respectively. Then, the objective is to calculate the Galois group of the equation

$$\frac{d}{dx}\begin{pmatrix} \xi_1 \\ \xi_2 \end{pmatrix} = A\begin{pmatrix} \xi_1 \\ \xi_2 \end{pmatrix}, \quad A \in Mat(2, K). \tag{2.5}$$

We also assume that the Newton polygon of the above equation has only one integer slope $k \in \mathbf{N}^*$. This is called the non-ramified case and the general case can be reduced to this one. By the Newton polygon of (2.5) we mean the Newton polygon of the equivalent second order single differential equation in $z = \frac{1}{x}$, i.e., the Newton polygon of the differential polynomial $P[D] = pD^2 + qD + r$ (the equation is $P[D]\xi = 0$), with $D = \frac{d}{dz}, p \in C[z], q, r \in C\{z\}$, $p(0) = q(0) = 0$. By the Fuchs theory, it is not difficult to see that the point $z_0 = \infty$ is an irregular singular point if, and only if, the Newton polygon has a side with slope in $(0, \infty)$.

By the classical theory (Huhukara-Turritin, [105]), there is a fundamental matrix U of (2.5), such that,

$$U = x^L e^Q H, \quad L \in M(2, C),$$

where $Q = \text{diag}(q_1, q_2)$, $q_1, q_2 \in C[x]$, $LQ = QL$, and H is holomorphic in any open angular sector at ∞ of opening angle $< \pi/k$,

$$S_d(\pi/k) := \{t : |x| > a, \arg x \in (d - \pi/2k, d + \pi/2k)\},$$

with a a suitable constant and $k := \deg(q_1 - q_2)$. Then H has an asymptotic expansion (whose entries are formal series):

$$H \sim \hat{H}, \quad \hat{H} \in GL(2, \hat{K}).$$

Here d is the argument of the bisecting line of the sector. A sector is characterized by d, α, where α is the opening. Then we will denote this sector by $S_d(\alpha)$.

This means that equation (2.5) has the formal solution

$$\hat{U} = x^L e^Q \hat{H}, \quad U \sim \hat{U}.$$

We observe that for $q_1 = q_2 = 0$, we are in the regular situation (the singular point is a singular regular one, and the formal series $\hat{U}$ is convergent).

In order to state the Ramis theorem we need some terminology: the exponential torus, the formal monodromy and the Stokes multipliers.

The exponential torus of (2.5) is defined (up to an isomorphism) as the Galois differential group of the Picard-Vessiot extension

$$\hat{K} \subset \hat{K}(e^{q_1}, e^{q_2}).$$

We see that this group is the Galois group of the trivial equation, considered over $\hat{K}$,

$$\frac{d\xi_i}{dx} = \frac{dq_i}{dx}\xi_i, \ i = 1, 2.$$

This exponential torus is (isomorphic to) C^* or $(C^*)^2$ if the rank of the $\mathbf{Z}$-module M_Q generated by $\{q_1, q_2\}$ is one or two, respectively. In the first case, the action of C^* is defined by

$$\lambda : e^{q_i} = e^{n_i s} \mapsto \lambda^{n_i} e^{n_i s}, \quad ,\lambda \in C^*, \quad \langle s \rangle = M_Q,$$

and, in the second case,

$$\lambda_i : e^{q_i} \mapsto \lambda_i e^{q_i}, \quad i = 1, 2.$$

By definition this action is constant on the coefficient field $\hat{K}$.

The formal monodromy is the transformation $\hat{M} \in GL(2, \mathbf{C})$, such that

$$\hat{U} \mapsto \hat{U}\hat{M},$$

when we formally make the circuit

$$x \mapsto e^{2\pi i}x.$$

It is clear that, by analytic continuation, it is possible to continue the analytic solution U to sectors $S_d(\alpha)$, with $\alpha > \pi/k$. The problem is that, in this new sector, this solution is not necessarily asymptotic to $\hat{U}$. The lines that bound the sectors where the asymptotic relation (2.5) remains valid are called Stokes rays. These lines are characterized by

$$\lim |x|^{-\deg(q_1-q_2)} Re(q_1 - q_2) = 0,$$

when $|x|$ tends to ∞ along this line. We can think that the analytic continuation from a sector $S_d(\pi/2)$, where the asymptotic expansion (2.5) is satisfied, is obtained by rotating the bisecting line d (in both directions), but preserving the opening π/k. Then we stop when a bounding side of the sector reaches a Stokes ray. The bisecting line d_s of this bounding sector $S_{d_s} := S_{d_s}(\pi/k)$ is called a singular line (sometimes it is called an anti-Stokes ray). They are characterized by the maximal exponential decay for $e^{q_1-q_2}$ or $e^{q_2-q_1}$. By the general theory, there are two sectors $S_{d_s+\epsilon}$, $S_{d_s-\epsilon}$, ($d_s + \epsilon$ means a small change in the argument of d_s by ϵ, $0 < \epsilon < \pi/2k$, and keeping the opening less than π/k). Hence, we get two analytical solutions U^+, U^- defined over S_{d_s} (by analytical continuation to this sector, ϵ going to 0). Then we have

$$U^- = U^+ Sto_{d_s},$$

where, by definition, the matrix $Sto_{d_s} \in GL(2, \mathbf{C})$ is the Stokes matrix in the singular direction d_s. It is possible to see that these Stokes matrices are unipotents, i.e., of the form (in the suitable fundamental system)

$$\begin{pmatrix} 1 & \lambda \\ 0 & 1 \end{pmatrix}, \quad \text{or,} \quad \begin{pmatrix} 1 & 0 \\ \mu & 1 \end{pmatrix}.$$

In particular, they belong to $SL(2, C)$. The complex numbers μ, λ are called the Stokes multipliers. In particular they belong to $SL(2, \mathbf{C})$.

   In an analogous, but more delicate, way (in this general case the phenomenon of multi-summability appears) we may describe the exponential torus, the formal monodromy and the Stokes matrices for a local system of differential equations of arbitrary dimension m

$$\frac{d\xi}{dx} = A\xi, \quad A \in Mat(m, K) \tag{2.6}$$

(see [74, 16]). Then

Theorem 2.3 ([90, 69, 16]) *The Galois (local) group of (2.6) is topologically generated by the exponential torus, the formal monodromy and the Stokes matrices (at $x = 0$).*

   We note that among these generators the main source of non-integrability comes from the Stokes multipliers. For example, it is not difficult to prove that the Zariski closure of the group (algebraically) generated by the two matrices

$$\begin{pmatrix} 1 & \lambda \\ 0 & 1 \end{pmatrix}, \quad \begin{pmatrix} 1 & 0 \\ \mu & 1 \end{pmatrix},$$

where λ, μ are both different from zero, is $SL(2, C)$ [18].
   It is possible to generalize the above theorem to a global linear differential equation in the following way. Let K be the field of meromorphic functions on a Riemann surface X and $S \subset X$ a discrete set. Then (see Appendix B).

Theorem 2.4 ([77]) *Let*

$$\frac{d\xi}{dx} = A\xi, \quad A \in Mat(m, K) \tag{2.7}$$

be a linear differential equation, S being the set of singular points (i.e., poles of the entries of A). Let P_i be the set of Stokes matrices and exponential torus at each of the singular points $a_i \in S$, and let M be the usual monodromy group of (2.7). Then the Galois group of (2.7) is topologically generated by P_i ($a_i \in S$) and M.

2.6   Coverings and differential Galois groups

In concrete differential equations it is useful, if possible, to replace the original differential equation over a compact Riemann surface, by a new differential equation over the Riemann sphere $\mathbf{P}^1$ (i.e., with rational coefficients) by a change of the independent variable. This equation on $\mathbf{P}^1$ is called the algebraic form of the equation. In a more general way we will consider the effect of a finite ramified covering on the Galois group of the original differential equation. In Appendix B the following theorem is proved.

Theorem 2.5 ([77]) *Let X be a (connected) Riemann surface. Let $f : X' \longrightarrow X$) be a finite ramified covering of X by a Riemann surface X'. Let ∇ be a meromorphic connection on X. We set $\nabla' = f^*\nabla$. Then we have a natural injective homomorphism*

$$\text{Gal}(\nabla') \to \text{Gal}(\nabla)$$

of differential Galois groups which induces an isomorphism between their Lie algebras.

We observe that, in terms of the differential Galois groups, this theorem means that the identity component of the differential Galois group is invariant by the covering.

An algebraic version of the above theorem is given by Katz [51]. This result is also proved in [8] (Proposition 4.7) for the particular case of a Fuchsian differential equation (see also [24, 25, 27, 9]). It is the mapping version for the so-called (in the cited references) method of reduction by discrete symmetries. Then this method is also valid in our more general setting. It is important to notice that, if one of the connections in the proposition is symplectic, then the identity components of the Galois groups of both connections are symplectic too.

2.7 Kovacic's algorithm

The Kovacic algorithm gives us a procedure in order to compute the Picard-Vessiot extension (i.e., a fundamental system of solutions) of a second order differential equation, provided the differential equation is integrable. Reciprocally, if the differential equation is non-integrable, the algorithm does not work (see[56]). In this (necessarily brief) description of the algorithm we essentially follow the version of the algorithm given in [33, 32]. The author is indebted to Anne Duval for some clarifications about his papers.

Given a second order linear differential equation with coefficients in $\mathbf{C}(x)$, it is a classical fact that it can be transformed to the so-called reduced invariant form

$$\xi'' - g\xi = 0, \tag{2.8}$$

with $g = g(x) \in \mathbf{C}(x)$.

We remark that in this change we introduce the exponentiation of a quadrature and the integrability of the original equation is equivalent to the integrability of the above equation although, in general, the Galois groups are not the same.

The algorithm is based on the following two general facts:

(A) The classification of the algebraic subgroups of $SL(2, \mathbf{C})$ given in Proposition 2.2 (the Galois group of the equation (2.8) is contained in $SL(2, \mathbf{C})$: see Section 2.1).

(B) The well-known transformation to a Riccati equation, by the change $v = \xi'/\xi$,

$$v' = g + v^2. \tag{2.9}$$

Then (see Section 2.2) the differential equation (2.8) is integrable, if and only if, the equation (2.9) has an algebraic solution. The *key* point now is that the degree n of the associated minimal polynomial $Q(v)$ (with coefficients in $\mathbf{C}(x)$) belong to the set

$$L_{\max} = \{1, 2, 4, 6, 12\}.$$

The determination of the set L of possible values for n, is the **First Step** of the algorithm. We remark that for $n = 4$, $n = 6$ and $n = 12$, the Galois group of (2.8) is finite (hence these values are related to the crystalographic groups). The two other steps of the algorithm (**Second Step** and **Third Step**) are devoted to computation of the polynomial $Q(v)$ (if it exists). If the algorithm does not work (i.e., if the equation (2.9) has no algebraic solution) then equation (2.8) is non-integrable and its Galois group is $SL(2, \mathbf{C})$.

Now we will describe the algorithm.

Let

$$g = g(x) = \frac{s(x)}{t(x)},$$

with $s(x)$, $t(x)$ relatively prime polynomials, and $t(x)$ monic. We define the following function h on the set $L_{\max} = \{1, 2, 4, 6, 12\}$, $h(1) = 1$, $h(2) = 4$, $h(4) = h(6) = h(12) = 12$.

First Step

If $t(x) = 1$ we put $m = 0$, else we factorize $t(x)$ in monic relatively prime polynomials. Then

1.1. Let Γ' be the set of roots of $t(x)$ (i.e., the singular points at the finite complex plane) and let $\Gamma = \Gamma' \cup \infty$ be the set of singular points. Then the order at a singular point $c \in \Gamma'$ is, as usual, $o(c) = i$ if c is a root of multiplicity i of $t(x)$. The order at infinity is defined by $o(\infty) = \max(0, 4 + \deg(s) - \deg(t))$. We call m^+ the maximum value of the order that appears at the singular points in Γ, and Γ_i is the set of singular points of order $i \leq m^+$.

1.2. If $m^+ \geq 2$ then we write $\gamma_2 = \operatorname{card}(\Gamma_2)$, else $\gamma_2 = 0$. Then we compute

$$\gamma = \gamma_2 + \operatorname{card}(\cup_{\substack{k \, odd \\ 3 \leq k \leq m^+}} \Gamma_k).$$

1.3. For the singular points of order one or two, $c \in \Gamma_2 \cup \Gamma_1$, we compute the principal parts of g:

$$g = \alpha_c(x - c)^{-2} + \beta_c(x - c) + O(1),$$

if $c \in \Gamma'$, and

$$g = \alpha_\infty x^{-2} + \beta_\infty x^{-3} + O(x^{-4}),$$

for the point at infinity.

1.4. We define the subset L' (of possible values for the degree of the minimal polynomial $Q(v)$) as $\{1\} \subset L'$ if $\gamma = \gamma_2$, $\{2\} \subset L'$ if $\gamma \geq 2$ and $\{4, 6, 12\} \subset L'$ if $m^+ \leq 2$.

1.5. We have the three following mutually exclusive cases:

1.5.1. If $m^+ > 2$, then $L = L'$.

1.5.2. If $m^+ \leq 2$ and the two following conditions are satisfied:

    1.5.2.1. For any $c \in \Gamma$, $\sqrt{1 + 4\alpha_c} \in \mathbf{Q}$, and $\sum_{c \in \Gamma'} \beta_c = 0$,

    1.5.2.2. For any $c \in \Gamma$ such that $\sqrt{1 + 4\alpha_c} \in \mathbf{Z}$, logarithmic term does not appear in the local solutions in a neighborhood of c,

then $L = L'$.

1.5.3. If cases 1.5.1 and 1.5.2 do not hold then $L = L' - \{4, 6, 12\}$.

1.6. If $L = \emptyset$, then equation (2.8) is non-integrable with Galois group $SL(2, \mathbf{C})$, else one writes n for the minimum value in L.

(We remark that Condition 1.5.2.2 is new. As the reader can check, it follows trivially from the fact that the existence of a logarithm in a local solution is an obstruction to have a finite monodromy and Galois group. I decided to include this condition here because it has been applied with success in some important applications [49].)

For the **Second Step** and the **Third Step** of the algorithm we consider the value of n fixed.

Second Step

2.1. If ∞ has order 0 we write the set

$$E_\infty = \{0, \frac{h(n)}{n}, 2\frac{h(n)}{n}, 3\frac{h(n)}{n}, \ldots, n\frac{h(n)}{n}\}.$$

2.2. If c has order 1, then $E_c = \{h(n)\}$.

2.3. If $n = 1$, for each c of order 2 we define

$$E_c = \{\frac{1}{2}(1 + \sqrt{1 + 4\alpha_c}, \frac{1}{2}(1 - \sqrt{1 + 4\alpha_c}\}.$$

2.4. If $n \geq 2$, for each c of order 2, we define

$$E_c = \mathbf{Z} \cap \{\frac{h(n)}{2}(1 - \sqrt{1 + 4\alpha_c}) + \frac{h(n)}{n}k\sqrt{1 + 4\alpha_c} : k = 0, 1, \ldots, n\}.$$

2.5. If $n = 1$, for each singular point of even order 2ν, with $\nu > 1$, we compute the numbers α_c and β_c defined (up to a sign) by the following conditions:

    2.5.1. If $c \in \Gamma'$,

$$g = \{\alpha_c(x - c)^{-\nu} + \sum_{i=2}^{\nu-1} \mu_{i,c}(x - c)^{-i}\}^2 + \beta_c(x - c)^{-\nu-1} + O((x - c)^{-\nu}),$$

and we write

$$\sqrt{g}_c := \alpha_c(x - c)^{-\nu} + \sum_{i=2}^{\nu-1} \mu_{i,c}(x - c)^{-i}.$$

    2.5.2. If $c = \infty$,

$$g = \{\alpha_\infty x^{\nu-2} + \sum_{i=0}^{\nu-3} \mu_{i,\infty} x^i\}^2 - \beta_\infty x^{\nu-3} + O(x^{\nu-4}),$$

and we write

$$\sqrt{g}_\infty := \alpha_\infty x^{\nu-2} + \sum_{i=0}^{\nu-3} \mu_{i,\infty} x^i.$$

Then for each c as above, we compute

$$E_c = \{\frac{1}{2}(\nu + \epsilon\frac{\beta_c}{\alpha_c}) : \epsilon = \pm 1\},$$

and the sign function on E_c is defined by

$$\text{sign}(\frac{1}{2}(\nu + \epsilon\frac{\beta_c}{\alpha_c})) = \epsilon,$$

being $+1$ if $\beta_c = 0$.

2.6. If $n = 2$, for each c of order ν, with $\nu \geq 3$, we write $E_c = \{\nu\}$.

Third Step

3.1. For n fixed, we try to obtain elements $\mathbf{e} = (e_c)_{c\in\Gamma}$ in the cartesian product $\prod_{c\in\Gamma} E_c$, such that:

(i) $d(\mathbf{e}) := n - \frac{n}{h(n)} \sum_{c\in\Gamma} e_c$ is a non-negative integer,

(ii) If $n = 2$ then there is at least one odd number in $\mathbf{e}$.

If no element $\mathbf{e}$ is obtained, we select the next value in L and go to the **Second Step**, else n is the maximum value in L and the Galois group is $SL(2, \mathbf{C})$ (i.e., the equation (2.8) is non-integrable).

3.2. For each family $\mathbf{e}$ as above, we try to obtain a rational function Q and a polynomial P, such that

(i)

$$Q = \frac{n}{h(n)} \sum_{c\in\Gamma'} \frac{e_c}{x - c} + \delta_{n1} \sum_{c\in\cup_{\nu>1}\Gamma_{2\nu}} \text{sign}(e_c)\sqrt{g_c},$$

where δ_{n1} is the Kronecker delta.

(ii) P is a polynomial of degree $d(\mathbf{e})$ and its coefficients are found as a solution of the (in general, overdetermined) system of equations

$$P_{-1} = 0,$$

$$P_{i-1} = -(P_i)' - QP_i - (n - i)(i + 1)gP_{i+1}, \quad n \geq i \geq 0,$$

$$P_n = -P.$$

If a pair (P, Q) as above is found, then equation (2.8) is integrable and the Riccati equation (2.9) has an algebraic solution v given by any root v of the equation

$$\sum_{i=0}^{n} \frac{P_i}{(n-1)!}v^i = 0.$$

If no pair as above is found we take the next value in L and we go to the **Second Step**. If n is the greatest value in L then equation (2.8) is non-integrable and the Galois group is $SL(2, \mathbf{C})$.

We notice that a remarkable simplification of the above algorithm was obtained in [100] for irreducible differential equations, and an algorithm for third order differential equations is given in [97, 98].

2.8 Examples

We now illustrate the Picard-Vessiot Theory with some examples. As we are interested to know when the identity component of the Galois group is abelian (see Chapter 4), we make it explicit in the known cases.

2.8.1 The hypergeometric equation

The hypergeometric (or Riemann) equation is the more general second order linear differential equation over the Riemann sphere with three regular singular singularities. If we place the singularities at $x = 0, 1, \infty$ it is given by

$$\frac{d^2\xi}{dx^2} + \left(\frac{1-\alpha-\alpha'}{x} + \frac{1-\gamma-\gamma'}{x-1}\right)\frac{d\xi}{dx}$$
$$+ \left(\frac{\alpha\alpha'}{x^2} + \frac{\gamma\gamma'}{(x-1)^2} + \frac{\beta\beta' - \alpha\alpha' - \gamma\gamma'}{x(x-1)}\right)\xi = 0, \qquad (2.10)$$

where (α, α'), (γ, γ'), (β, β') are the exponents at the singular points and must satisfy the Fuchs relation $\alpha + \alpha' + \gamma + \gamma' + \beta + \beta' = 1$. We denote the exponent differences by $\hat{\lambda} = \alpha - \alpha'$, $\hat{\nu} = \gamma - \gamma'$ and $\hat{\mu} = \beta - \beta'$.

We also use one of its reduced forms

$$\frac{d^2\xi}{dx^2} + \frac{c - (a+b+1)x}{x(x-1)}\frac{d\xi}{dx} - \frac{ab}{x(x-1)}\xi = 0, \qquad (2.11)$$

where a, b, c are parameters, with the exponent differences $\hat{\lambda} = 1-c$, $\hat{\nu} = c-a-b$ and $\hat{\mu} = b - a$, respectively.

Now, we recall the theorem of Kimura that gives necessary and sufficient conditions for the hypergeometric equation to have integrability.

Theorem 2.6 ([52]) *The identity component of the Galois group of the hypergeometric equation (2.10) is solvable if, and only if, either*

(i) *at least one of the four numbers $\hat{\lambda}+\hat{\mu}+\hat{\nu}$, $-\hat{\lambda}+\hat{\mu}+\hat{\nu}$, $\hat{\lambda}-\hat{\mu}+\hat{\nu}$, $\hat{\lambda}+\hat{\mu}-\hat{\nu}$ is an odd integer, or*

(ii) *the numbers $\hat{\lambda}$ or $-\hat{\lambda}$, $\hat{\mu}$ or $-\hat{\mu}$ and $\hat{\nu}$ or $-\hat{\nu}$ belong (in an arbitrary order) to some of the following fifteen families*

1	$1/2 + l$	$1/2 + m$	arbitrary complex number	
2	$1/2 + l$	$1/3 + m$	$1/3 + q$	
3	$2/3 + l$	$1/3 + m$	$1/3 + q$	$l + m + q$ even
4	$1/2 + l$	$1/3 + m$	$1/4 + q$	
5	$2/3 + l$	$1/4 + m$	$1/4 + q$	$l + m + q$ even
6	$1/2 + l$	$1/3 + m$	$1/5 + q$	
7	$2/5 + l$	$1/3 + m$	$1/3 + q$	$l + m + q$ even
8	$2/3 + l$	$1/5 + m$	$1/5 + q$	$l + m + q$ even
9	$1/2 + l$	$2/5 + m$	$1/5 + q$	$l + m + q$ even
10	$3/5 + l$	$1/3 + m$	$1/5 + q$	$l + m + q$ even
11	$2/5 + l$	$2/5 + m$	$2/5 + q$	$l + m + q$ even
12	$2/3 + l$	$1/3 + m$	$1/5 + q$	$l + m + q$ even
13	$4/5 + l$	$1/5 + m$	$1/5 + q$	$l + m + q$ even
14	$1/2 + l$	$2/5 + m$	$1/3 + q$	$l + m + q$ even
15	$3/5 + l$	$2/5 + m$	$1/3 + q$	$l + m + q$ even

Here l, m and q are integers.

We recall that Schwarz's table gives us the cases for which the Galois (and monodromy) groups are finite (i.e., the identity component of the Galois group is reduced to the identity element) and is given by fifteen families. These families are given by families 2–15 of the table above and by the family $(1/2 + \mathbf{Z}) \times (1/2 + \mathbf{Z}) \times \mathbf{Q}$ (see, for instance, [88]). As this last family is already contained in family 1 of the above table, all of the Schwarz's families are, of course, contained in the above table.

2.8.2   The Bessel equation

The Bessel equation is

$$x^2 \frac{d^2\xi}{dx^2} + x\frac{d\xi}{dx} + (x^2 - n^2)\xi = 0, \tag{2.12}$$

with n a complex parameter. This equation is a particular confluent hypergeometric equation (by a limit process two of the singular points in a variant of the hypergeometric equation coincide).

As (2.12) is one of the most simple but non-trivial (i.e., in general, non-integrable) equations with Stokes phenomenon, we are going to make explicit for it the concepts introduced in Section 2.5.

First, we observe that the Galois group is contained in $SL(2, \mathbf{C})$, since $1/x$ is a logarithmic derivative (see Section 2.2). It is an equation with two singular points, 0, ∞, the first one being regular singular and the second one irregular. We are interested in the point at infinity.

There are several ways to compute the matrices Q and L of Section 2.5. For example, we can follow the general constructive method of the Huhukara-Turritin theory [105, 10]. First, we make a formal change

$$\begin{pmatrix} \xi \\ \xi' \end{pmatrix} = \hat{P} \begin{pmatrix} u_1 \\ u_2 \end{pmatrix},$$

where $P \in Mat(2, \hat{K})$ ($\hat{K} := \mathbf{C}[[x^{-1}]][x]$) which formally diagonalizes the equation. The solution is precisely the formal solution in equation (2.12), and is found step by step in a recursive way ([105, 10]). In this way we obtain $q_1 = ix = -q_2$ and $L = -1/2I$. The exponential torus is $\mathbf{C}^*$ and the formal monodromy $\hat{M} = -I$.

The Stokes rays are $\mathbf{R}_+$ and $\mathbf{R}_-$, and the singular lines $i\mathbf{R}_+$, $i\mathbf{R}_-$. Hence, we have two Stokes multipliers (one for each singular line),

$$St_1 = \begin{pmatrix} 1 & \mu \\ 0 & 1 \end{pmatrix},$$

$$St_2 = \begin{pmatrix} 1 & 0 \\ \lambda & 1 \end{pmatrix}.$$

But, for this equation the global theory (coefficients in $\mathbf{C}(x)$) and the local one (coefficients in $K = \mathbf{C}\{\{x^{-1}\}\}[x]$) are essentially the same. We note that the actual monodromy M_0 around 0 and around ∞ are the same, therefore the differential Galois group at the origin can be interpreted as a subgroup of the differential Galois group at infinity. It is possible to compute the actual monodromy M_0 in the classical basis at the origin, which is of course different from the basis at infinity introduced in the previous computation. We get $M_0 = \mathrm{diag}(e^{2\pi in}, e^{-2\pi in})$.

It is easy to relate the actual monodromy and the formal monodromy at infinity using the Stokes multipliers:

$$M_0 = St_1 \hat{M} St_2.$$

Now, as the trace is an invariant, we get

$$\mathrm{tr}\, M_0 = 2\cos(2\pi n) = -\lambda\mu - 2, \ \lambda\mu = -4\cos^2 \pi n.$$

Hence, if n does not belong to $\mathbf{Z} + 1/2$, the Bessel equation is non-integrable. In fact, this necessary condition for integrability is also sufficient. So by the classical theory (see, for example, [62]) it is well known that the Bessel functions for $n \in \mathbf{Z} + 1/2$ can be expressed by elementary functions: the Picard-Vessiot extension is obtained by exponentiation of integrals of elements of $\mathbf{C}(x)$.

2.8.3   The confluent hypergeometric equation

One of the forms of the general confluent hypergeometric equation is given by
the Whittaker equation [109]

$$\frac{d^2\xi}{dz^2} - (\frac{1}{4} - \frac{\kappa}{z} + \frac{4\mu^2 - 1}{4z^2})\eta = 0, \qquad (2.13)$$

with parameters κ and μ. The singular points are $z = 0$ (regular) and $z = \infty$
(irregular).

As in the case of the Bessel equation we have two singular lines associated
to the irregular point for (2.13). For computation of the Galois group, the
following proposition is useful ([69], Subsection 3.3))

Proposition 2.5 *There is a fundamental system of solutions such that if α, β
are the two Stokes multipliers corresponding to the two singular lines, with
corresponding Stokes matrices*

$$\begin{pmatrix} 1 & \alpha \\ 0 & 1 \end{pmatrix},$$

$$\begin{pmatrix} 1 & 0 \\ \beta & 1 \end{pmatrix},$$

then
  (i) *$\alpha = 0$ if, and only if, either $\kappa - \mu \in \frac{1}{2} + \mathbf{N}$ or $\kappa + \mu \in \frac{1}{2} + \mathbf{N}$.*
  (ii) *$\beta = 0$ if, and only if, either $-\kappa - \mu \in \frac{1}{2} + \mathbf{N}$ or $-\kappa + \mu \in \frac{1}{2} + \mathbf{N}$.*
  *Furthermore (with respect to the same fundamental system of solutions),
  the group generated by the formal monodromy and the exponential torus
  is given by the multiplicative group*

$$\left\{ \begin{pmatrix} \delta & 0 \\ 0 & \delta^{-1} \end{pmatrix} : \delta \in \mathbf{C}^* \right\}.$$

As a consequence, we get the abelian criterion expressed in terms of the
parameters $p := \kappa + \mu - \frac{1}{2}$ and $q := \kappa - \mu - \frac{1}{2}$.

Corollary 2.1 *The identity component G^0 of (2.13) is abelian if, and only if,
(p, q) belong to $(\mathbf{N} \times (-\mathbf{N}^*)) \cup ((-\mathbf{N}^*) \times \mathbf{N})$ (i.e., p, q are integers, one of them
being positive and the other negative).*

We observe that the abelian case (G^0 abelian) for the Whittaker equation
is only possible when the two Stokes multipliers are zero and this corresponds to
the diagonal case 4 of the classification given by Proposition 2.2 (the Whittaker
equation is a symplectic one). If only one of the Stokes multipliers is different

from zero we are in case 6 of this classification, and we have integrability but the identity component of the Galois group is not abelian. If two of the Stokes multipliers are different from zero, we fall in case 7 with Galois group $SL(2, \mathbf{C})$, as we remarked in Section 2.5.

If in the Bessel equation (2.12) we make the change of the dependent variable $\xi = x^{-1/2}\eta$ and of the independent variable $x = z/2i$, we get a Whittaker equation

$$\frac{d^2\eta}{dz^2} - (\frac{1}{4} + \frac{4n^2 - 1}{4z^2})\eta = 0, \tag{2.14}$$

with parameters $\kappa = 0$ and $\mu = n$. As in the above change we only introduce algebraic functions, the identity component of the Galois group of the Bessel equation is preserved.

2.8.4   The Lamé equation

The algebraic form of the Lamé Equation is [88, 109]

$$\frac{d^2\eta}{dx^2} + \frac{f'(x)}{2f(x)}\frac{d\eta}{dx} - \frac{Ax + B}{f(x)}\eta = 0, \tag{2.15}$$

where $f(x) = 4x^3 - g_2 x - g_3$, with A, B, g_2 and g_3 parameters such that the discriminant of f, $27g_3^2 - g_2^3$ is non-zero. This equation is a Fuchsian differential equation with four singular points over the Riemann sphere.

With the well-known change $x = \mathcal{P}(z)$, we get the Weierstrass form of the Lamé equation

$$\frac{d^2\eta}{dz^2} - (A\mathcal{P}(z) + B)\eta = 0, \tag{2.16}$$

where $\mathcal{P}$ is the elliptic Weierstrass function with invariants g_2, g_3 (we recall that $\mathcal{P}(z)$ is a solution of the differential equation $(\frac{dx}{dz})^2 = f(x)$). Classically the equation is written with the parameter n instead of A, with $A = n(n+1)$. This equation is defined on a torus Π (a genus one Riemann surface) with only one singular point at the origin. Let $2w_1$, $2w_3$ be the two periods of the Weierstrass function $\mathcal{P}$ and $\mathbf{g}_1$, $\mathbf{g}_2$ their corresponding monodromies in the above equation. If $\mathbf{g}_*$ represents the monodromy around the singular point, then $\mathbf{g}_* = [\mathbf{g}_1, \mathbf{g}_2]$ ([109, 88]).

By the above theorem we know that the identity component of the Galois group is preserved by the covering $\Pi \to \mathbf{P}^1$, $t \mapsto x$.

The relation between the monodromy groups of equations (2.15) and (2.16) is discussed in [88], Chapter IX. From a modern point of view it is studied in [27].

Now we study the integrability of the Lamé equation (2.16) which is equivalent to the integrability of (2.15). In fact, we have a stronger result: the identity components of the Galois groups of both equations are isomorphic (by Theorem 2.5).

First, it is easy to see that a necessary and sufficient condition for the *total* Galois group of (2.16) to be abelian is that $n \in \mathbf{Z}$. We sketch the steps of the proof. Indeed, this is a classical well-known necessary and sufficient condition for the monodromy group M of the equation (2.16) to be abelian (it is clear that, as M is generated by $\mathbf{g}_1$ and $\mathbf{g}_2$, an equivalent condition for the abelianess of M is $\mathbf{g}_* = \mathbf{1}(identity)$, and the indicial equation at the singularity is $\rho^2 - \rho - n(n+1) = 0$, and there is no logarithmic term for n integer. Therefore, as G is topologically generated by M, it must also be abelian.

Now the known (mutually exclusive) cases of closed form solutions of the Lamé equation (2.16) are as follows:

(i) The Lamé and Hermite solutions [34, 42, 88, 109]. In this case n is an arbitrary integer and the three other parameters are arbitrary. In the case of the Lamé solutions there is one solution that is an elliptic function with the same periods as the function $\mathcal{P}$ (i.e., it belongs to the coefficient field K), hence, by the normality of the Picard-Vessiot extensions, in this case the Galois group of the equation (2.16) is of type 3 of Proposition 2.2. We will use this property in the last chapter.

(ii) The Brioschi-Halphen-Crawford solutions [7, 34, 42, 88]. Now $m := n + \frac{1}{2} \in \mathbf{N}$ and the parameters B, g_2 and g_3 must satisfy an algebraic equation

$$0 = Q_m(g_2/4, g_3/4, B) \in \mathbf{Z}[g_2/4, g_3/4, B],$$

where Q_m has degree m in B. This polynomial is known as the Brioschi determinant and we will construct it later in this section.

(iii) The Baldassarri solutions [7]. The condition on n is $n + \frac{1}{2} \in \frac{1}{3}\mathbf{Z} \cup \frac{1}{4}\mathbf{Z} \cup \frac{1}{5}\mathbf{Z} - \mathbf{Z}$, with additional (involved) algebraic restrictions on the other parameters.

We notice that, by the above arguments, case (i) exhaust all the possible abelian cases for the Galois group G of equation (2.16) (i.e., types 1 abelian, 2, 3 with $k = 1, 2$ and 4 in Proposition 2.2). Furthermore cases (ii) and (iii) exhaust all the other possibilities of purely algebraic solutions (i.e., G finite). In other words, the known solutions cover types 1, 2, 3 with $k = 1, 2$ and 4 of Proposition 2.2. We are left now with type 3 (with $k > 2$), type 5 and type 6, to complete the study of the integrability of equation (2.16).

Proposition 2.6 ([81]) *The equation* (2.16) *is integrable only in the cases (i), (ii) and (iii) above.*

Proof. For type 5 of Proposition 2.2, by Proposition 2.3 the associated Riccati equation, $v' = g + v^2$, $g(z) = -(n(n+1)\mathcal{P}(z) + B)$, must have two solutions, $v_{1,2} = \kappa \pm \sqrt{\omega}$, in a quadratic extension of $K = \mathcal{M}(\Pi)$ (field of meromorphic functions on the Riemann surface Π of genus one). Therefore $\kappa, \omega \in K$ satisfy the system

$$\kappa' = \kappa^2 + \omega + g,$$

$$\omega' = 4\kappa\omega.$$

These equations are found in [7] and in what follows we use some of the methods of this paper.

The above system is equivalent to

$$\frac{1}{4}\left(\frac{\omega'}{\omega}\right)' - \frac{1}{16}\left(\frac{\omega'}{\omega}\right)^2 - \omega = g \qquad (2.17)$$

(this equation was well known in the classical literature, see [43], p. 35).

If $v_i = -\xi_i'/\xi_i$, $i = 1, 2$, proceeding as in the proof of Proposition 2.2 we get $w^2/(4\omega) = \xi_1^2\xi_2^2$. On the other hand let α_1, α_2 be a fundamental system of solutions corresponding to the indicial equation around the singular point $z = 0$ (modulo periods). That is, $\alpha_1 = z^{\rho_1}\phi_1(z)$, $\alpha_2 = z^{\rho_2}\phi_2(z)$ ($\phi_1(0) \neq 0$, $\phi_2(0) \neq 0$), where $\rho_1 = n+1$, $\rho_2 = -n$ are the roots of the indicial equations at the origin and logarithmic terms can not appear, because $w^2/(4\omega) = \xi_1^2\xi_2^2$ and $\omega \in K$.

Expressing ξ_1, ξ_2 as linear combinations of α_1, α_2 we obtain

$$\frac{w^2}{4\omega} = az^{4m+4}\phi_1^4 + bz^{2m+3}\phi_1^3\phi_2 + cz^2\phi_1^2\phi_2^2$$

$$+dz^{-2m+1}\phi_1\phi_2^3 + ez^{-4m}\phi_2^4.$$

Furthermore, $\xi_1^2\xi_2^2$ is an elliptic function whose only pole is $z = 0$ (modulo periods), because this is the only singular point of the solutions of the Lamé equation. Therefore $w^2/(4\omega) = z^{-k}\phi$, $k \in \mathbf{N}$, $\phi(0) \neq 0$, where ϕ is a holomorphic function in a neighborhood of $z = 0$. We have the following mutually exclusive possibilities:

(a) $2m \in \mathbf{Z}$ (and $m \neq \mathbf{Z}$, otherwise we fall into the Lamé or Hermite solutions), else

(b) $4m \in \mathbf{Z}$, $2m \in \mathbf{Z}$.

Case (a) corresponds to the Brioschi-Halphen-Crawford solutions since those are the only ones such that $2m \in \mathbf{Z}$, $m \neq \mathbf{Z}$ and they have no logarithmic term.

In case (b), $b = d = 0$ and we can take $m > 0$ because if $4m \leq -5$ then $4m + 4 \leq -1$, and the values $4m = -1$, $4m = -3$ are excluded because $w^2/4\omega$

must have a pole at $z = 0$. Therefore $4m = k \in \mathbf{N}$, k odd. Then $\omega = w^2/(4\xi_1^2\xi_2^2)$ is an elliptic function of odd order k, having a zero of order k in $z = 0$ and hence a pole of odd order at some point $z = z_0 \neq 0$ (module periods). On the other hand $(\omega'/\omega)^2$ and $(\omega'/\omega)'$ are elliptic functions with double poles at $z = 0$ and at the poles of ω. From (2.17) it follows that g has a pole at $z_0 \neq 0$ contradicting the fact that it has only one pole at $z = 0$ (module periods). Type 5 does not occur in the equation (2.16).

The impossibility of type 3 (with $k > 2$) and of type 6 of Proposition 2.2 is simpler. By a direct computation, the derived group G' is given by unipotent triangular matrices. But as the local monodromy around the singular point $\mathbf{g}_* \in G'$, the exponents (i.e., solutions of the indicial equation) must be integers and we are in the Lamé or Hermite solutions with an abelian Galois group, contradicting the assumption that the Galois group is of type 3 (with $k > 2$) or type 6. We have finished the proof.   □

We will need two more results about equation (2.16). The first one is very elementary, we state it as a proposition for future references.

Proposition 2.7 *Assume that for equation (2.16) we have* $\mathbf{g}_1^2 = 1$ *(or* $\mathbf{g}_2^2 = 1$*),* $\mathbf{g}_i$, $i = 1, 2$, *being the monodromies along the periods. Then the Galois group of this equation is abelian.*

Proof. From $\mathbf{g}_1^2 = 1$ it follows that $\mathbf{g}_1 = 1$ or $\mathbf{g}_1 = -1$ (because $\mathbf{g}_1$ is in $SL(2, \mathbf{C})$). If $\mathbf{g}_1 = 1$, it is clear that $\mathbf{g}_* = [\mathbf{g}_1, \mathbf{g}_2] = 1$ (the case $\mathbf{g}_1 = -1$ is analogous).   □

The second result is not so elementary and we need some terminology.

We recall that the moduli of the elliptic curve $v^2 = 4u^2 - g_2 - g_3$ (we write the elliptic curve in the canonical form, where as above g_2 and g_3 are the invariants) is characterized by the value of the modular function j,

$$j = j(g_2, g_3) = \frac{g_2^3}{g_2^3 - 27g_3^2}. \tag{2.18}$$

We recall that two elliptic curves are birationally equivalent if, and only if, they have the same value of the modular function (see, for instance [93]).

Although the conditions on g_2, g_3 and B for a finite Galois group (case (iii)) are difficult to systematize, there is, in this case, a general result by Dwork answering a question posed by Baldassarri in [7].

Proposition 2.8 ([35]) *Assume that the Galois group of equation (2.16) is finite. Then for a fixed value of n, the number of possible couples (j, B) is finite.*

We note that the proof of Dwork is given for the algebraic form of the Lamé equation (equation (2.15)). But as by a finite covering the identity component of the Galois group is preserved (Theorem 2.5), then the finiteness of the Galois group of equation (2.15) is equivalent to the finiteness of the Galois group of equation (2.16) (a linear algebraic group is finite if, and only if, its identity component is trivial) and the result is valid also for equation (2.16).

The author is indebted to B. Dwork for sending him the above result.

Finally, for the families of type (ii) we recall the computation of the Brioschi determinant following Baldassarri [7] (it will be important in the applications of Chapter 6). If in the Lamé equation we make the Halphen substitution [42] $z = 2\hat{z}$ and use the addition theorem for $\mathcal{P}$ (see [109]) we obtain

$$\frac{d^2\xi}{d\hat{z}^2} - 4\left[n(n+1)\left(\frac{1}{4}\left(\frac{\mathcal{P}''(\hat{z})}{\mathcal{P}'(\hat{z})}\right)^2 - 2\mathcal{P}(\hat{z})\right) + B\right]\xi = 0. \tag{2.19}$$

If $(2\omega_1, 2\omega_3)$ are the periods of $\mathcal{P}$, the singularity of (2.16) at $z = 0$ (modulo the periods) is transformed to the singularities of (2.19)

$$\hat{z} = 0, \omega_1, \omega_2, \omega_3$$

(modulo the periods), where $\omega_1 + \omega_2 + \omega_3 = 0$. Now, to complete the Halphen transformation, we perform the change $\xi = \left(\mathcal{P}'(\hat{z})\right)^{-n} w$, obtaining

$$\frac{d^2 w}{d\hat{z}^2} - 2n\frac{\mathcal{P}''(\hat{z})}{\mathcal{P}'(\hat{z})}\frac{dw}{d\hat{z}} + 4\left(n(2n-1)\mathcal{P}(\hat{z}) - B\right)w = 0,$$

with singularities as above. Now let $x = \mathcal{P}(\hat{z})$ be a *new* independent variable (i.e., we have a finite covering $z \mapsto x$). We get the following algebraic form for the above equation

$$\left(-x^3 + \frac{g_2}{4}x + \frac{g_3}{4}\right)\frac{d^2 w}{dx^2} + \left(3x^2 - \frac{g_2}{4}\right)(m-1)\frac{dw}{dx} + \left[B - (2m-1)(m-1)x\right]w = 0, \tag{2.20}$$

having singularities at ∞, e_1, e_2, e_3 (corresponding to the previous ones

$$0, \omega_1, \omega_2, \omega_2$$

in (2.19)). We recall that $m = n + \frac{1}{2}$.

The exponents associated to the singularities are $(0, m)$ at $e_i, i = 1, 2, 3$, and $(-2m + 1, -m + 1)$ at ∞. As the difference is $m \in \mathbf{N}$ there will appear, in general, logarithmic terms. But if in one of the singularities there are no logarithmic terms they do not appear in any of the other singularities, because all the singularities come from the unique singularity of (2.16) by means of the Halphen transformation. Furthermore, if in an equation over $\mathbf{P}^1$ all the exponents are integers and there are no logarithmic terms, then the general solution is rational. In particular, if this happens in (2.20), then we have integrability for the Lamé equation.

To avoid logarithmic terms at $x = \infty$, a necessary and sufficient condition is the existence of a Laurent series solution of the form

$$w = \sum_{j=0}^{\infty} c_j\, x^{2m-j-1}\ ,\quad c_0 \neq 0, \tag{2.21}$$

corresponding to the lower exponent $-2m + 1$.

This leads to a recurrent system for the coefficients $c_0, c_1, \ldots$, which, in particular, gives the uncoupled system:

$$
\begin{array}{llll}
B\,c_0 & + \quad (m-1)c_1 & & =0, \\
(2m-1)(m-1)\tfrac{g_2}{4}c_0 & + \quad B\ \ c_1 & +2(m-2)c_2 & =0, \\
(2m-1)(2m-2)\tfrac{g_3}{4}c_0 & +(2m-2)(m-2)\tfrac{g_2}{4}c_1+B\ c_2+3(m-3)c_3 & & =0, \\
(2m-2)(2m-3)\tfrac{g_3}{4}c_1 & +(2m-3)(m-3)\tfrac{g_2}{4}c_2+B\ c_3+4(m-4)c_4 & & =0, \\
\end{array}
$$

$$\vdots$$
$$\vdots$$

$$
\begin{array}{lll}
(m+3)(m+2)\tfrac{g_3}{4}c_{m-4}+(m+2)2\tfrac{g_2}{4}c_{m-3} & +B\,c_{m-2}+(m-1)1c_{m-1}=0, \\
(m+2)(m+1)\tfrac{g_3}{4}c_{m-3}+(m+1)1\tfrac{g_2}{4}c_{m-2} & +B\,c_{m-1} & =0.
\end{array}
$$

Therefore, the necessary and sufficient condition to have a solution of the form (2.21) is

$$Q_m\left(\frac{g_2}{4}, \frac{g_3}{4}, B\right) = 0, \tag{2.22}$$

where $Q_m\left(\frac{g_2}{4}, \frac{g_3}{4}, B\right)$ is the determinant of dimension m of the coefficients of the above linear system in the variables $c_0, c_1, \ldots, c_{m-1}$. This is the Brioschi determinant.

We observe that all the examples in this section are second order differential equations over the Riemann sphere (in the case of the Lamé equation we consider its algebraic form), then it is theoretically possible to apply the Kovacic's algorithm [56, 33].

Chapter 3

Hamiltonian Systems

This chapter is devoted to explaining some concepts and results on Hamiltonian systems. We focus our attention on the concept of complete integrability i.e., Liouville integrability: the existence of n independent first integrals in involution, n being the number of degrees of freedom. Although integrability is well defined for these systems, it is very important to clarify what kind of regularity is allowed for the first integrals: differentiability or analyticity in the real situation, analytic, meromorphic or algebraic (meromorphic and meromorphic at infinity) first integrals in the complex setting.

Another important problem is to clarify the dynamical meaning of the different concepts of integrability or non-integrability. For instance, it is clear in the differential real situation that an open dense domain of the phase space is foliated by Lagrangian invariant manifolds. In the opposite direction one can ask for the dynamical meaning of non-integrability. In fact, it is well known that, in the real context, the existence of transversal homoclinic orbits is an obstruction to the integrability by real analytical first integrals. But in the complex situation, practically all is unknown about the global dynamical behaviour of non-integrable systems. In Chapter 7, for a particular situation, we will make connection between the purely algebraic differential Galois obstruction to integrability of Chapter 4 and the existence of transversal homoclinic orbits.

In this chapter we assume, unless otherwise stated, that we are either in the real (differentiable or analytical) situation or in the complex analytic situation. So all the objects that we consider (manifolds, functions, tensors, morphisms, bundles, etc...) are endowed with the same degree of regularity, respectively. We denote the ground numerical field by k. So, $k = \mathbf{R}$ in the real case and $k = \mathbf{C}$ in the complex case.

This chapter can not be considered as an introduction to Hamiltonian systems. As in Chapter 2, we have, in general, presented only the required

J. J. Morales Ruiz, *Differential Galois Theory and Non-Integrability of Hamiltonian*, Systems, Modern Birkhäuser Classics, DOI: 10.1007/978-3-0348-0723-4_3, © Springer Basel 1999

definitions and results without proofs. On the other hand, we prove the results of Section 3.4 about the abelian structure of some subalgebras of the Poisson algebra of rational functions. These results were obtained as part of joint work of the author with J.P. Ramis [77] and they are essential preliminaries for proof of the fundamental theorems of Chapter 4.

3.1   Definitions

We remark that in classical books (see, for instance, [4, 3, 65]), definitions are given in the real (differentiable or analytical) context, but all of them work well in the complex (holomorphic) setting. But in order to be able to deal with complex singularities, we give some specific definitions valid in the complex situation only (we will follow [77]).

A symplectic manifold is a manifold M endowed with a closed, non-degenerate two-form Ω: the symplectic form. It is well known that the dimension of M over k must be even, $2n$. The symplectic manifolds are the phase space of Hamiltonian systems and their tangent bundles are symplectic bundles which have already been defined in Section 2.3 in the complex setting (for real symplectic bundles, see [65]).

Example. Let $Q = k^n$ and $M = T^*Q \approx k^{2n}$ be its cotangent bundle. We get a symplectic structure on M defining the symplectic form $\Omega = \sum_{i=1}^n dy_i \wedge dx_i = d(\sum_{i=1}^n y_i dx_i)$, with x_i being the coordinate functions on the base Q and y_i the coordinates on the fiber T_x^*Q. It is clear that the matrix of Ω has in this case the canonical form J of Section 2.3 (in a formal way one can say that there is a canonical or symplectic frame in the symplectic bundle TM)

$$J = \begin{pmatrix} 0 & I \\ -I & 0 \end{pmatrix}.$$

This example can be easily generalized to an arbitrary manifold Q. The coordinates x_i are the positions and y_i are the canonically conjugated momenta. The space Q is called the configuration space.

The above example gives us the general structure of the symplectic form in local coordinates according to the Darboux theorem.

Theorem 3.1 (Darboux Theorem) *In a neighborhood of each point p of the symplectic manifold M there is a system of coordinates (x_i, y_i) (canonical coordinates) such that the symplectic form is expressed in the canonical form* $\Omega = \sum_{i=1}^n dy_i \wedge dx_i$.

As in Riemannian geometry, the non-degeneracy of the symplectic form defines the musical isomorphism of vector bundles

$$\flat : TM \to T^*M,$$

given by $\Omega(X,) = \flat X$, with X a vector field. As is usual in the theory of vector bundles we identify the vector bundle with its sections. This can also be written as the contraction $i_X \Omega = \flat \cdot X$. We remark that in canonical coordinates (i.e., in a symplectic frame in TM and in the dual frame in T^*M) $\flat^{-1}$ is given by the matrix J.

A Hamiltonian vector field is a vector field X_H defined on the symplectic manifold M such that $X_H = \flat^{-1} \cdot dH$, where H is a function: the Hamilton function or simply the Hamiltonian. The differential equations of the integral curves of a Hamiltonian vector field X_H are called a Hamiltonian system and as a direct consequence of the Darboux theorem it is possible to write them as

$$\dot{x}_i = \frac{\partial H}{\partial y_i}, \qquad \dot{y}_i = -\frac{\partial H}{\partial x_i},$$

$i = 1, \dots, n$. The above system of equations is called the Hamilton equations associated to the Hamiltonian H. Sometimes we will write the above equations in a more compact form as

$$\dot{z} = X_H(z),$$

with $z = (x_1, \dots, x_n, y_1, \dots, y_n)$. Also one often uses the same name for the Hamiltonian system and for the vector field X_H.

Among the several invariants by a Hamiltonian flow, the most important one is the symplectic form Ω. In other words, $L_{X_H}\Omega = 0$, where we write L_X for the Lie derivative with respect to X, and one says that the flow is symplectic. As we will see in Chapter 3, a direct consequence of the above is that the variational equations along a particular solution are given by symplectic connections. In particular, the Galois group of the variational equations must be symplectic.

As the Hamiltonian flow is symplectic, the powerful tool of canonical transformations can be used to simplify the system; ideally we would like to solve it completely. A canonical transformation is a change of variables from the (canonical) variables (x_i, y_i) and the time t to the variables $(\hat{x}_i, \hat{y}_i)$, $\hat{t}$ ($i = 1, 2, \dots, n$), such that the "extended" symplectic form $\sum_{i=1}^{n} dy_i \wedge dx_i - dH \wedge dt$ remains invariant. So, $\sum_{i=1}^{n} dy_i \wedge dx_i - dH \wedge dt = \sum_{i=1}^{n} d\hat{y}_i \wedge d\hat{x}_i - dH \wedge d\hat{t}$. Then the Hamilton equations are also invariant by the transformation. We note that the above definition is more general than the usual one in which the symplectic form remains invariant.

By means of the musical isomorphism $\flat$ it is possible to endow the regular functions on M with a special structure of Lie algebra: a Poisson algebra. It is clear that the application $f \mapsto X_f$ between the functions and the Hamiltonian vector fields is k-linear. In order to obtain a homomorphism of Lie algebras, we have to define a Lie bracket over the functions $\{,\}$, the Poisson bracket, such that

$$X_{d\{f,g\}} = [X_f, X_g],$$

for given functions f and g. Hence, the Poisson bracket is given by

$$\{f, g\} = \Omega(X_f, X_g).$$

In canonical coordinates it has the classical expression

$$\{f, g\} = \sum_{i=1}^{n} \frac{\partial f}{\partial y_i} \frac{\partial g}{\partial x_i} - \frac{\partial f}{\partial x_i} \frac{\partial g}{\partial y_i}.$$

Then the algebra of functions (differentiable or analytical, in the real case; holomorphic or meromorphic in the complex case) has the following important properties.

Proposition 3.1 *The functions over M endowed with the Poisson bracket have the structure of a Poisson algebra. In other words, they satisfy the properties*

(i) *they are a Lie algebra (with the Bracket $\{,\}$),*

(ii) $\{h, fg\} = f\{h, g\} + g\{h, f\}$ *(Leibniz rule).*

We note that the set of functions over M have two different algebraic structures: it is a commutative algebra with respect to the ordinary product, and it is a Poisson algebra (in particular, a Lie algebra). The property (ii) above is a compatibility condition between these two structures.

Furthermore, given a Hamiltonian vector field X_H we have the following

Proposition 3.2 (1) *The function f is a first integral of X_H if, and only if, $\{H, f\} = 0$ (we say that the functions f and H are in involution or commute). In particular, H is always a first integral of X_H.*

(2) *The set of first integrals of X_H is itself a Poisson subalgebra with the Poisson bracket (Jacobi's theorem).*

(3) *If the functions f and g are in involution, then the flows of the Hamiltonian fields X_f and X_g must commute, i.e., $\phi_f \circ \phi_g = \phi_g \circ \phi_f$, if ϕ_f, ϕ_g are the flows of f and g, respectively.*

We note that (2) follows directly from (1) and from the Jacobi identity.

It is important to recall that in the two above propositions we can consider several degrees of regularity for the algebra of functions over M. So, it is true for the algebra of differentiable or analytical functions in the real case and for the algebra of holomorphic or meromorphic functions in the complex case. Moreover, it is even possible to consider more regular cases, for instance the algebra of polynomials or the algebra of rational functions (if the symplectic manifold M is a vector space of dimension $2n$).

Now we look at the meromorphic case. Let H be a given holomorphic function over M and X_H the associated holomorphic Hamiltonian vector field. For some applications (cf., introduction of the *points at infinity* in Chapter 4 below) it is necessary to allow *meromorphic* Hamiltonian functions. The corresponding Hamiltonian vector field will be of course also meromorphic. But it can also be necessary to allow degeneracy points or poles for the canonical form Ω. Then we consider a complex connected manifold M' of complex dimension $2n$, endowed with a closed meromorphic 2-form Ω. We suppose that Ω is holomorphic and regular over a not void open subset $M \subset M'$. Then we can choose M such that $M' - M$ is an analytic, non-necessarily regular, hypersurface $M_\infty \subset M'$. The form Ω is a meromorphic section of the bundle $\Lambda^2 T^* M$. Therefore it is equivalent to say that the tangent bundle TM admits a structure of a meromorphic symplectic bundle. The manifold $(M, \Omega_{|M})$ is clearly a symplectic manifold. We call M_∞ the hypersurface at infinity.

Example. Let $M' = \mathbf{P}^1 \times \mathbf{P}^1$, $M = \mathbf{C}^2$, $M_\infty = \{\infty\} \times \mathbf{P}^1 \cup \mathbf{P}^1 \times \{\infty\}$. We denote $(x, y) \in \mathbf{C}^2$ and by x', y' the coordinates at infinity over respectively the first and second factor $\mathbf{P}^1$ of M'. Then we set $\Omega = dx \wedge dy$ over M. It extends uniquely to a meromorphic form over M' and we have

$$\Omega = \frac{dx' \wedge dy'}{x'^2 y'^2}$$

over a neighborhood of $\{\infty\} \times \{\infty\}$.

We go back to our general situation.

To a holomorphic (resp., meromorphic) vector field X over M', we can associate a *meromorphic* 1-form α, using the formula

$$\alpha(Y) = \Omega(Y, X).$$

We get the usual musical map $\flat$ by restriction to M. This map is an isomorphism over M. Writing the application $\flat$ in coordinates in a neighborhood of a point at infinity we see that $\flat$ admits an inverse. This inverse is holomorphic over M but can have poles over M_∞.

We finish this section with two remarks. So far we have only considered autonomous Hamiltonian systems (Hamiltonian vector fields independent of time). It is possible to generalize the above structures to the non-autonomous situation (see Section 5.1 of [3]). From an intrinsic point of view this is done in the context of contact geometry, instead of symplectic geometry, which, in coordinates, is given by the extended symplectic form

$$\sum_{i=1}^{n} dy_i \wedge dx_i - dh \wedge dt,$$

defined in the $2n+1$-dimensional extended phase space parametrized by (x_i, y_i, t) $(i = 1, \ldots, n)$. Then the Hamilton equations are given by the same expression, with the Hamilton function depending on time. We note that for linear non-autonomous Hamiltonian systems an alternative intrinsic description has already been given in Chapter 2 of this monograph in the context of symplectic connections.

We remark that usually in classical mechanics the Hamiltonian H is the energy of the system, and for a classical Hamiltonian system it is given by

$$H = T + V,$$

where T (the kinetic energy) is a definite positive quadratic form with respect to the momenta y_i (with coefficients, in general, depending on the coordinates x_i) and V (the potential) is a function of the coordinates x_i, $i = 1, 2, \ldots, n$ only. Frequently people define a Hamiltonian system by its potential, $V = V(x_1, x_2, \ldots, x_n)$, then it is assumed that the kinetic energy is given by the typical expression in Euclidean coordinates

$$T = \frac{1}{2} \sum_{i=1}^{n} y_i^2.$$

We will also follow these conventions.

3.2   Complete integrability

Let X_H be a Hamiltonian system defined over a symplectic manifold M of dimension $2n$. One says that a Hamiltonian system X_H is completely integrable or Liouville integrable if there are n functions $f_1 = H$, $f_2, \ldots, f_n$, such that

(1) They are functionally independent i.e., the 1-forms df_i $i = 1, 2, \ldots, n$, are linearly independent over a dense open set $U \subset \mathcal{M}$, $\bar{U} = M$.

(2) They form an involutive set, $\{f_i, f_j\} = 0$, $i, j = 1, 2, \ldots, n$.

We remark that by (2) the functions f_i, $i = 1, \ldots, n$ are first integrals. An important property is the following.

Proposition 3.3 Let M be a symplectic manifold of dimension $2n$. Let

$$f_1, \ldots, f_{n+1}$$

be an involutive set of functions. Then the functions $f_1, \ldots, f_n, f_{n+1}$ are functionally dependent.

This proposition is well known in the real case. In the complex case the same proof works well. As the proof is very simple and instructive we shall give it (for $k = \mathbf{C}$) for the sake of completeness.

Proof. We can interpret $\flat$ as an isomorphism between the holomorphic fiber bundles TM (tangent bundle) and T^*M (cotangent bundle). We denote by $\natural$ the inverse isomorphism.

We assume that the functions $f_1, \ldots, f_n, f_{n+1}$ are functionally independent. Then the $\mathbf{C}$-linear forms $df_1(x), \ldots, df_{n+1}(x)$ are linearly independent for every $x \in U$ (U is a dense open set in M). Let $x_0 \in U$. We set $f_i(x_0) = c_i \in \mathbf{C}$. The subset $\Sigma = \{f_1 = c_1, \ldots, f_{n+1} = c_{n+1}\}$ is an analytic (smooth) *submanifold* of complex dimension $n-1$. The vector fields $Y_i = \natural df_i$ ($i = 1, \ldots, n+1$) are tangent to Σ ($df_i(Y_j) = \{f_i, f_j\} = 0$) and linearly independent over the complex field at each point of U. (The linear map $\natural$ induces an isomorphism between T_xM^* and T_xM.) This implies $\dim \Sigma \geq n + 1$ and we get a contradiction. $\square$

There are several possible generalizations of the above definition of integrability for Hamiltonian systems. An apparently more general definition of complete integrability for a Hamiltonian system X_H is obtained considering the n functionally independent first integrals in involution f_i, $i = 1, \ldots, n$, without the assumption that one of them is the Hamiltonian H itself. But then by the last proposition, the Hamiltonian is functionally dependent on $f_1, \ldots, f_n$ and the set given by $H, f_2, \ldots, f_n$ satisfies the two properties above. Therefore the two definitions are equivalent.

One of the most elementary cases of (complete) integrability is the following.

Example. Let X_H be the Hamiltonian system of n degrees of freedom, defined in canonical coordinates by the Hamiltonian

$$H = H_1 + H_2 + \cdots + H_n,$$

where each $H_i = H_i(x_i, y_i)$ i.e., a function of x_i, y_i only. Then the Hamiltonian system X_H is completely integrable, H_i being, $i = 1, \ldots, n$ the first integrals in involution. In this case we say that X_H is separable. In fact this is not the most general definition of separability, but this is the only one that we will use.

It is clear that the concept of complete integrability, as defined above, deals with the existence of an involutive Poisson subalgebra of the Poisson algebra of first integrals of X_H. Indeed, the vector space generated by $f_1, \ldots, f_n$ (over the numerical field k) is an involutive Poisson algebra. In particular, we get an abelian Lie algebra, and by the duality given by the symplectic structure (the musical isomorphism $\flat^{-1}$) one obtains the abelian Lie algebra generated by $X_{f_1}, \ldots, X_{f_n}$, as well as an abelian Lie group isomorphic to k^n. This consideration is, in the real case, the starting point in the proof of the "geometrical" part of the Liouville theorem.

Theorem 3.2 (Liouville Theorem) *Let X_H be a real completely integrable Hamiltonian system defined over a real symplectic manifold M of dimension $2n$. Let $f_1, \ldots, f_n$ be a set of involutive first integrals and $M_{\mathbf{a}} = \{\mathbf{z} \in M : f_i(z) = a_i, \ i = 1, \ldots, n\}$, $\mathbf{a} = (a_1, \ldots, a_n)$, a non-critical level manifold of $f_1, \ldots, f_n$ (that is, $\mathrm{rank}(df_1, \ldots, df_n) = n$ over $M_{\mathbf{a}}$). Then*

    (a) $M_{\mathbf{a}}$ is an invariant manifold by the flow of the Hamiltonian system X_H, and if it is compact and connected it is diffeomorphic to an n-dimensional torus $T^n = \mathbf{R}^n / \mathbf{Z}^n$ (Liouville torus).

    (b) In a neighborhood of the torus $M_{\mathbf{a}}$ there is a canonical system of coordinates $(\mathbf{I}, \phi) = (I_1, \ldots, I_n, \phi_1, \ldots, \phi_n)$, $\phi_i (\mathrm{mod}\, 2\pi)$, the action-angle coordinates, such that the action coordinates $\mathbf{I}$ correspond to a transversal direction to the torus and the angle variables ϕ are the coordinates on the torus. Then, Hamilton's equations for X_H expressed in these coordinates are

$$\dot{I}_i = 0,$$

$$\dot{\phi}_i = \omega_i,$$

$i = 1, \ldots, n$, where the frequencies ω_i depend on the action variables only, i.e., $\omega_i = \omega_i(I_1, \ldots, I_n)$, and the Hamiltonian system of Hamiltonian H is integrated by quadratures.

    Before some remarks about the Liouville theorem it is convenient to introduce a definition. A Lagrangian manifold in a (real or complex) $2n$-dimensional symplectic manifold M is an n-dimensional (differentiable or analytical, respectively) submanifold L of M such that the symplectic form Ω restricted to L is identically zero. A reference on Lagrangian manifolds is [107].

    Although we do not give here a proof of the Liouville theorem (a complete proof is given in [4]), we make some comments about the method used in the proof.

    As remarked, the geometrical part of the theorem ((a) above), is based on a particular realization of the Frobenius theorem: the distribution X_{f_i}, $i = 1, \ldots, n$ with integral manifolds $M_{\mathbf{a}}$ is not only involutive but also abelian. This gives an action of the Lie group $\mathbf{R}^n$ (of the Lie algebra generated by

X_{f_i}, $i = 1, \ldots, n$) on the manifold $M_{\mathbf{a}}$. It is proved then that the above action is transitive and the isotropy group is a discrete subgroup of $\mathbf{R}^n$ and hence isomorphic to $\mathbf{Z}^n$. From the general theory of homogeneous spaces, claim (a) follows.

The computational part of the Liouville theorem ((b) above), is based on the Hamilton-Jacobi theory of canonical transformations. The action coordinates are essentially given by the n periods of the action differential 1-form $\sum_{i=1}^{n} y_i dx_i$ along the n-torus $M_{\mathbf{a}}$, and the angle coordinates are the natural coordinates on the torus. In order to solve explicitly in a closed way a given completely integrable Hamiltonian system, in the original coordinates $(x_1, \ldots, x_n, y_1, \ldots, y_n)$, it is necessary to make quadratures and inversion of differentiable functions. Although this program is in general a difficult task, it is theoretically possible. The dynamics of the system is known: an open dense domain U of the phase space is foliated by invariant manifolds and the flow restricted to them is linear. We observe that the invariant manifolds $M_{\mathbf{a}}$ are Lagrangian manifolds, because if $\mathbf{a}$ is fixed, the action coordinates I_i, $i = 1, \ldots, n$, are constant and are moment type coordinates, such that the symplectic form is expressed as $\Omega = \sum_{i=1}^{n} dI_i \wedge d\phi_i$.

In the general complex setting with holomorphic or meromorphic first integrals, instead of differentiable ones, it seems difficult to extend the Liouville theorem. As in the real case, the phase space is foliated by Lagrangian manifolds, but unfortunately, it seems that a description of the general topology of these manifolds and of the flow of X_H on them is missing, for the most general case.

However, in the algebraic (complex) context and under some additional assumptions, normally associated to the existence of a "good" Lax pair for the system, or the solvability of the system by abelian functions, it is possible also to get invariant complex manifolds (roughly speaking abelian varieties) such that the flow on them is linear, and the general solution of the system is expressed by abelian functions. These are the so-called algebraically integrable systems (see, for instance [1, 2, 84, 103]). It is remarkable that most of the known non-trivial completely integrable Hamiltonian systems are solved by this method, but there are completely integrable systems that are not algebraically integrable. For instance, the trivial 1-degree of freedom classical Hamiltonian system with Hamiltonian

$$H = \frac{1}{2} y^2 + V(x),$$

where the potential V is a polynomial of degree greater than 4 or less than 3, is not algebraically integrable. We note that if the degree of V is 3 or 4, the abelian variety is an elliptic curve, i.e., a complex torus.

Despite the last remark, in the applications, we are interested in complex analytical (holomorphic or meromorphic) Hamiltonian systems such that when the coordinates are real then the Hamiltonian system is also real. In other words, we obtain the real Hamiltonian system as a subsystem, when we restrict the time and the dependent variables to real values. In this case we can apply the Liouville theorem to the real system, and then we can make the analytical continuation, i.e., we simply consider the variables of the system, including time, as complex.

These considerations justify the use of the definition of complete integrability also in the complex setting (with holomorphic or meromorphic first integrals in involution). From now on integrability always means complete integrability.

We observe the analogy between this concept of integrability for Hamiltonian systems and the concept of integrability for linear differential equations given in Chapter 2 in the context of differential Galois theory. The Liouville extensions of Chapter 2 correspond to the Liouville theorem here. For the identity component of the Galois group solvable (equivalently, the Lie algebra of the Galois group is solvable) there, we have here an abelian Lie algebra of symmetries of the Hamiltonian system. We remark that it is interesting and not casual that the name of Liouville is attached to both concepts of integrability. In fact Liouville devoted an important part of his life to the search for a general theory of integrability for differential equations.

As a last remark and although this monograph is not oriented in this direction (for this reason we do not enter in the technical details and definitions, but we will give references), we would like to point out another relation, between differential Galois theory and integrability of Hamiltonian systems which seems not well-studied. Concretely, there is a direct connection between algebraically completely integrable systems and the strongly normal extensions of Kolchin (see [54] and [15], Chapter IV, Section 3). It would be interesting to clarify completely this connection. For instance, we will state the following question.

Question. *What is the relation between the singularity theory of Adler and van Moerbeke on algebraic completely integrable Hamiltonian systems (see [2, 103]) and singularity theory of the Kolchin's strongly normal extensions as studied by Buium ([15], Chapter IV)?*

3.3　Three non-integrability theorems

In this section we shall explain three non-integrability results for Hamiltonian systems: a theorem proved by Poincaré for real Hamiltonian systems, a theorem proved by Ziglin for complex Hamiltonian systems and a theorem proved by Lerman for real Hamiltonian systems. All of them are based upon a study of the variational equations (VE) along particular integral curves.

Let X_H be a real Hamiltonian system defined on a (real) symplectic manifold M of dimension $2n$,

$$\dot{z} = X_H(z),$$

and let Γ be a periodic orbit of this system of period T given by $z = z(t)$, $z(t+T) = z(t)$.

We recall that the variational equations (VE) of a Hamiltonian system X_H along the integral curve Γ are defined by the linear non-autonomous differential system of equations

$$\dot{\xi} = X'_H(z(t))\xi.$$

We set $A = A(t) = X'_H(z(t))$. Let $U(t)$ be a fundamental system of solutions of the VE along Γ, with $U(0) = I$ (identity matrix), then $U(t)$ is the linear part of the flow $\phi(t)$ of X_H along Γ, i.e.,

$$\phi'(t) \cdot \xi = U(t) \cdot \xi,$$

$\xi \in T_pM$, $p \in \Gamma$. The matrix $M = U(T)$ is called the monodromy matrix of the periodic orbit Γ and its eigenvalues are called multipliers of the periodic orbit Γ. As M is symplectic, $M \in Sp(n, \mathbf{R})$ and if λ is a multiplier then λ^{-1} must be also a multiplier.

Theorem 3.3 (Poincaré Theorem [87]) *Assume that a Hamiltonian system X_H has an involutive set $f_1 = H, \ldots, f_k$ of k first integrals ($k \leq n$) functionally independent on the periodic orbit Γ. Then at least $2k$ multipliers of Γ must be equal to one.*

Between the several equivalent proofs we select the more instructive for our ends (this proof will be intimately related with the process of reduction from the (VE) to the (NVE): see Section 4.1).

Sketch of the Proof. By restriction of the k Hamiltonian fields $X_{f_1}, \ldots, X_{f_k}$ to Γ, we obtain k independent solutions of the VE $v_i = X_{f_i}(z(t))$, $i = 1, \ldots, k$. For each point $p \in \Gamma$, we can take a symplectic base in the symplectic space T_pM that contains the k vectors v_i. In this base we have $2k$ eigenvectors with eigenvalues equal to one. $\qquad \square$

Corollary 3.1 *Let Γ be a periodic orbit of a completely integrable Hamiltonian system X_H, $f_1 = H, \ldots, f_n$ being an involutive set of first integrals independent over Γ. Then the $2n$ multipliers of Γ are equal to one.*

It is clear in the corollary that the periodic orbit, Γ, is contained in a Liouville torus (if the invariant manifold of the Liouville theorem that contains Γ is compact).

We remark that Poincaré's theorem can be generalized to the complex holomorphic setting (with the same proof) if instead of a periodic orbit we consider an element of the fundamental group of the Riemann surface defined by a complex integral curve, and instead of the monodromy of a periodic orbit we consider an element of the monodromy group of the variational equation. By one more step we do the same thing for the Galois group. In fact all of the above is a consequence of the process of reduction by k degrees of freedom from the variational equations (VE) to the normal variational equations (NVE), if there is an involutive set of k first integrals independent over the integral curve. This will be done in detail in Chapter 4 from the differential Galois point of view.

We note also the analogy of the method of reduction with the so-called d'Alambert reduction of the order for a scalar linear differential equation when some independent particular solutions are known (see, for instance, [46], p. 121–122). In the complex case, both methods must be considered as part of the same theory: the differential Galois theory.

In the complex situation, Ziglin in 1982 [114] proved a non-integrability theorem. He uses the constraints imposed by the existence of a sufficient number of first integrals on the monodromy group of the normal variational equations along some integral curve. This is a result about branching of solutions: the monodromy group expresses the ramification of the solutions of the normal variational equations in the complex domain.

We consider a complex analytic symplectic manifold of dimension $2n$ and a holomorphic Hamiltonian system X_H defined on it. Let Γ be the Riemann surface representing an integral curve $z = z(t)$ which is not an equilibrium point of the vector field X_H. Then we can write, as above, the variational equations (VE) along Γ,

$$\dot\eta = X'_H(z(t))\eta.$$

In general if, including the Hamiltonian, there are k analytical first integrals *independent over* Γ and in *involution*, we can reduce the number of degrees of freedom of the VE by k and get the normal variational equation (NVE) that, in suitable coordinates, can be written as a linear Hamiltonian system

$$\dot\xi = JS(t)\xi,$$

where, as usual,

$$J = \begin{pmatrix} 0 & I \\ -I & 0 \end{pmatrix}$$

is the matrix of the symplectic form of dimension $2(n - k)$.

As the NVE is a Hamiltonian linear system, its monodromy group is contained in the symplectic group. We note that the monodromy group is contained in the Galois group and the Galois group is symplectic (see Section 2.4 or Appendix C).

Theorem 3.4 (Ziglin) *Suppose that an n-degrees of freedom Hamiltonian system has $n - k$ additional meromorphic first integrals, independent of the above integrals over a neighborhood of Γ (but not necessarily over Γ itself) and assume also that the monodromy group of the NVE contains a non-resonant transformation g. Then, any other element of the monodromy group of the NVE sends eigendirections of g into eigendirections of g.*

We recall that a linear transformation of $Sp(m, \mathbf{C})$ is resonant if there are integers $r_1, \ldots, r_m$, such that $\lambda_1^{r_1} \cdots \lambda^{r_m} = 1$, with λ_i being its eigenvalues.

We remark that Ziglin's theorem *does not assume complete integrability* of the Hamiltonian system X_H. The n independent first integrals are of two types: k of them that are in involution and *independent over* Γ are used to make the reduction to the NVE only; the other $n - k$ first integrals, not necessarily in involution, restrict the possible structure of the monodromy group of the normal variational equation. However for two degrees of freedom ($n = 2$ in Ziglin's theorem) the Hamiltonian system is completely integrable.

Now the variational equation along a particular integral curve $z(t)$ of X_H always has the particular solution $X_H(z(t) = \dot{z}(t)$. We then get the normal variational equations by reducing one degree of freedom. This reduction is connected to a Poincaré map in the following way.

Let Γ be a periodic orbit of a *real* Hamiltonian system X_H with n-degrees of freedom (or a loop in the fundamental group of the Riemann surface that represents Γ, for the complex case). Then we consider a symplectic submanifold S (Poincaré section) of dimension $2n - 2$ transversal to Γ and contained in the energy manifold level that contains Γ, i.e., $S \subset H^{-1}(h)$ with $H(\Gamma) = h$. Then the Poincaré map P of the periodic orbit Γ (of a loop of the fundamental group of the Riemann surface Γ, in the complex setting) is a map defined in a neighborhood $U_p \subset S$ of a point $p \in \Gamma$ given by the intersection of the flow of the Hamiltonian system X_H with the section S. Then the linear part of the Poincaré map, $P'(p)$, is given precisely by the monodromy matrix of the normal variational equations (an element of the monodromy group of the normal variational equations, respectively, in the complex case).

If the integral curve Γ is not a periodic orbit, it is also possible to define a Poincaré map P between two transversal sections S_1 and S_2 (in a given energy level) to the integral curve Γ. As above, this map is defined by the flow of X_H and, in general, it is only defined in a neighborhood of the curve Γ.

We shall now restrict our attention to the real case in order to explain the theorem of Lerman. Let X_H be a two degrees of freedom *real* analytic

Hamiltonian system with a saddle-center equilibrium point $o \in M$, with $H(o) = 0$. The following two facts are well known:

(i) Associated to the "saddle part" are stable and unstable integral curves of X_H asymptotic to the equilibrium point.

(ii) Associated to the "center part" for sufficiently small energies h is a one-parameter family of periodic orbits (Liapounov orbits) with parameter h. The set of these orbits defines the center manifold of the equilibrium point.

Moreover, the above Liapounov orbits are unstable, i.e., the multipliers of the Poincaré map of these orbits are outside the unit circle. Hence each Liapounov orbit has two-dimensional stable and unstable invariant manifolds.

We make the additional assumption that there is a homoclinic orbit (or homoclinic loop), i.e., the continuation of the stable and unstable curves in (i) coincide in a single integral curve.

The key point now is to know if the stable and unstable invariant manifolds of the Liapounov orbits are the same or not. The importance of this fact is that if they are not the same, then, as a consequence of their intersection, the system has transversal homoclinic orbits, we have unpredictable chaotic dynamics (for instance, the Bernoulli shift is included as a subsystem) and the Hamiltonian system has no additional (real) analytical first integrals besides the Hamiltonian (see [83]).

Let Φ be the flow map along the homoclinic orbit between two points in it and contained in a small enough neighborhood U of the equilibrium point. Then

Theorem 3.5 ([64]) *There are suitable coordinates in U such that in these coordinates, the linearized flow*

$$d\Phi = \begin{pmatrix} K & Q \\ S & R \end{pmatrix},$$

where R is the 2×2 matrix corresponding to the normal variational equation along the homoclinic orbit. Now assume that the stable and unstable invariant manifolds of every (small enough) Lyapounov orbit are the same. Then R must be a rotation.

We remark that the matrix R gives us the linear part, dP, of the Poincaré map P (in some coordinates) along the homoclinic orbit between transversal sections through two points in it, and contained in a small enough neighborhood of the equilibrium point o.

Now if we asssume that the Hamiltonian is a classical one, $H = 1/2(y_1^2 + y_2^2) + V(x_1, x_2)$, and if the homoclinic orbit is contained in an invariant plane

(x_1, y_1), we can write the Hamiltonian as

$$H = \frac{1}{2}(y_1^2 + y_2^2) + \varphi(x_1) + \frac{1}{2}\alpha(x_1)x_2^2 + h.o.t.(x_2), \qquad (3.1)$$

where

$$\varphi(x_1) = -\frac{1}{2}\nu^2 x_1^2 + h.o.t.(x_1), \quad \alpha(x_1) = \omega^2 + h.o.t.(x_1), \qquad (3.2)$$

with ν and ω non-vanishing real parameters.

The above theorem is generalized to more degrees of freedom in [55], although we do not need this result here.

We shall see later in this monograph that the above three theorems, the Poincaré theorem (as it is interpreted above within the process of reduction from the VE to the NVE), the Ziglin theorem (with $n = k+1$, i.e., when X_H is completely integrable) and the Lerman theorem are connected in a deep way. This unity is transparent in the framework of the differential Galois theory.

3.4 Some properties of Poisson algebras

In this section we expose some results on certain properties of the Poisson algebra of rational functions over a symplectic space. These results were obtained in [77].

Let V be a symplectic complex space of (complex) dimension $2n$. We choose a symplectic basis $\{e_1, \ldots, e_n; \varepsilon_1, \ldots, \varepsilon_n\}$ and we denote by $(x_1, \ldots, x_n; y_1, \ldots, y_n)$ the coordinates of $v \in E$ in this basis. If $\langle v, v' \rangle$ is the symplectic product of $v, v' \in E$, then

$$\langle v, v' \rangle = \sum_{i=1}^{n} x_i y_i' - x_i' y_i.$$

We set

$$\mathbf{C}[V] = \bigoplus_{k \, 0} S^k V^*.$$

We endow $\mathbf{C}[V]$ with the ordinary multiplication. We get the commutative $\mathbf{C}$-algebra of polynomials on V. We denote by $\mathbf{C}(V)$ the field of fractions of $\mathbf{C}[V]$. Using the Poisson product, we endow $\mathbf{C}(V)$ with a structure of non-commutative $\mathbf{C}$-algebra, the Poisson structure.

Using coordinates we can compute the Poisson product of $f, g \in \mathbf{C}(V)$:

$$\{f, g\} = \sum_{i=1}^{n} \frac{\partial f}{\partial y_i}\frac{\partial g}{\partial x_i} - \frac{\partial f}{\partial x_i}\frac{\partial g}{\partial y_i}.$$

The two products on $\mathbf{C}(V)$ are related by the Leibniz rule

$$\{fg, h\} = f\{g, h\} + g\{f, h\}.$$

Let $A \subset \mathbf{C}(V)$ be a complex vector subspace. If it is stable by the Poisson product, then A is a Poisson subalgebra of $\mathbf{C}(V)$. The field of fractions of A is also a Poisson subalgebra of $\mathbf{C}(V)$.

Let A be an involutive subset of $\mathbf{C}(V)$. Then the complex vector subspace generated by A is also involutive and a Poisson subalgebra. Using the Leibniz rule we verify that the subalgebra (for the ordinary product) generated by A is involutive and is also a Poisson subalgebra.

A subset of an involutive subset is clearly also an involutive subset.

Let A be a subset of $\mathbf{C}(V)$. We define the orthogonal $A^{\perp}$ of A in $\mathbf{C}(V)$ by

$$A^{\perp} = \{f \in \mathbf{C}(V)/\{f, g\} = 0 \text{ for every } g \in A\}.$$

The biorthogonal of A is $A^{\perp\perp} = (A^{\perp})^{\perp}$.

Using the Leibniz rule and the Jacobi identity we verify immediately that $A^{\perp}$ is a subalgebra and a Poisson subalgebra of $\mathbf{C}(V)$. Therefore $A^{\perp\perp}$ is also a subalgebra and a Poisson subalgebra of $\mathbf{C}(V)$.

Let A be a subset of $\mathbf{C}[V]$. It is involutive if, and only if, we have the inclusion $A \subset A^{\perp}$. Moreover, if A is involutive, we have the inclusions

$$A \subset A^{\perp\perp} \subset A^{\perp}$$

and $A^{\perp\perp}$ is an involutive subalgebra of $\mathbf{C}(V)$.

Let A be a subalgebra of $\mathbf{C}(V)$ (for the ordinary product). We say that $f \in \mathbf{C}(V)$ is algebraic on A if there exists a non-trivial polynomial $P \in A[Z]$ such that $P(f) = 0$. The algebraic closure $\bar{A}$ of A in $\mathbf{C}(V)$ is by definition the set of the $f \in \mathbf{C}(V)$ that are algebraic on A. The algebraic closure $\bar{A}$ of A in $\mathbf{C}(V)$ is a subfield (it is the algebraic closure of the field of fractions of A). The following proposition and its corollary are essential in this book.

Proposition 3.4 *Let $A \subset \mathbf{C}(V)$ be an involutive subalgebra, i.e., a subalgebra for the ordinary product which is also an involutive subset. Let $\bar{A} \subset \mathbf{C}[V]$ be the algebraic closure of A in $\mathbf{C}(V)$. Then we have inclusions*

$$A \subset \bar{A} \subset A^{\perp\perp} \subset A^{\perp}$$

and $\bar{A}$ is an involutive subalgebra of $\mathbf{C}(V)$

Proof. Let $\beta \in \bar{A}$. Let $P \in A[Z]$ be a minimal polynomial for β.

We choose $\beta' \in A^{\perp}$. From $P(\beta) = 0$, we get $\{P(\beta), \beta'\} = 0$. Using $\beta \in \bar{A}$ and the Leibniz rule we see easily that the operator $\{., \beta'\}$ is A-linear derivation on $A[\beta]$, therefore

$$\{P(\beta), \beta'\} = \frac{\partial P}{\partial Z}(\beta)\{\beta, \beta'\} = 0.$$

As the polynomial P is minimal, we have $\frac{\partial P}{\partial Z}(\beta) \neq 0$ and therefore $\{\beta, \beta'\} = 0$. This yields $\bar{A} \subset A^{\perp\perp}$. As $A^{\perp\perp}$ is involutive, the subalgebra $\bar{A}$ is also involutive.
□

Corollary 3.2 *Let V be a symplectic complex space of dimension $2n$. Let $A \subset C(V)$ be a subalgebra (for the ordinary product) which is generated by a finite involutive subset $\alpha = \{\alpha_1, \ldots, \alpha_n\}$. We suppose that the n elements $\alpha_1, \ldots, \alpha_n$ are algebraically independent. Then*
  (i) *A is an involutive subalgebra,*
 (ii) *$A^{\perp}$ is an involutive subalgebra,*
(iii) *$\bar{A} = A^{\perp} = A^{\perp\perp}$*

Proof. Claim *(i)* is evident.

Let $f \in A^{\perp}$. It is orthogonal to $\alpha \subset A$. The n elements $\alpha_1, \ldots, \alpha_n$ are algebraically independent and in involution, therefore f and $\alpha_1, \ldots, \alpha_n$ are algebraically dependent. We admit this claim and we prove it later. Then $f \in \bar{A}$. We get an inclusion $A^{\perp} \subset \bar{A}$. Using the proposition, we get also the inclusions $\bar{A} \subset A^{\perp\perp} \subset A^{\perp}$. Therefore $\bar{A} = A^{\perp} = A^{\perp\perp}$. The subalgebra $A^{\perp\perp}$ is involutive. Claim *(ii)* follows.
□

Now we set as usual

$$J = \begin{pmatrix} 0 & I \\ -I & 0 \end{pmatrix},$$

where I is the identity matrix of dimension n. Then we have $J^t = -J = J^{-1}$.

A square matrix M of order $2n$ is symplectic if and only if

$$M^t J M = J.$$

A square matrix M of order $2n$ is in the Lie algebra of the Lie group of symplectic matrices if and only if

$$M^t J + J M = 0.$$

This is equivalent to $(JM)^t = JM$, i.e., to the fact that the matrix JM is symmetric. This Lie algebra is the set of linear Hamiltonian vector fields (with constant coefficients).

The symplectic structure on the symplectic vector space V gives the musical isomorphism

$$\flat : V \to V^*.$$

If X is the column vector of the coordinates of $v \in V$ in the chosen symplectic basis, then the column vector of the coordinates of $\flat(v)$ in the dual basis is JX.

Using the contraction $V^* \otimes V \to \mathbf{C}$ between the first and the third factor in $V^* \otimes V \otimes V$, we get a homomorphism

$$V^* \otimes V \to \operatorname{End}(V).$$

It is well known that this is an isomorphism and in general we will identify $V^* \otimes V$ and $\operatorname{End}(V)$ modulo this isomorphism.

We set $\psi = id_{V^*} \otimes (\frac{1}{2}\flat)$. The map $\psi : V^* \otimes V \to V^* \otimes V^*$ is an isomorphism. We can interpret ψ as an isomorphism $\operatorname{End}(V) \to V^* \otimes V^*$. An element $u \in \operatorname{End}(V)$ belongs to the Lie algebra $sp(V)$ if and only if $\psi(u)$ is invariant by the symmetry

$$V^* \otimes V^* \to V^* \otimes V^*$$

$$v \otimes w \mapsto w \otimes v.$$

Then $\psi(u)$ defines an element of $S^2 V^*$ and ψ induces an isomorphism

$$\phi : sp(V) \to S^2 V^*.$$

If as before we use the chosen symplectic basis of V and the dual basis of V^* and if we denote, respectively, by M and M' the matrices of $u \in sp(V)$ and the matrix of the quadratic form corresponding to $\phi(u) \in S^2 E^*$, then

$$M' = \frac{1}{2} JM.$$

We define an operation of the Lie algebra $\operatorname{End}(V)$ on V^* by $w \mapsto -u^t(w)$. Using this operation and the natural operation of $\operatorname{End}(V)$ on V, we get an operation of $\operatorname{End}(V)$ on $V^* \otimes V$:

$$w \otimes v \mapsto -u^t(w) \otimes v + w \otimes u(v).$$

If we identify $V^* \otimes V$ with $\operatorname{End}(V)$, then the corresponding operator is $[u, .]$.

Lemma 3.1 *(i) By the isomorphism $\phi : sp(V) \to S^2 V^*$, the operation of $sp(V)$ on V^* defined above corresponds to the action of $S^2 V^*$ on V^* given by the Poisson product.*

(ii) By the isomorphisms $\phi : sp(V) \to S^2 V^*$ and $\flat : V \to V^*$, the natural operation of $sp(V)$ on V corresponds to the action of $S^2 V^*$ on V^* by the Poisson product.

Proof. We use a canonical basis on V and the dual basis on V^*. Let M be the matrix of $u \in sp(V)$. Then $M' = \frac{1}{2}JM$ is a symmetric matrix. It is the matrix of the quadratic form corresponding to $\phi(u)$. We denote by X (resp. X') the column vector of the coordinates of $v \in V^*$ (resp. $v' \in V^*$) in the dual basis. If $X^t = (x; y)$ and $X'^t = (x'; y')$. If $f(x; y) = X^t M' X$ and $g(x; y) = X'^t X$. We denote $M' = \frac{1}{2}\begin{pmatrix} A & B \\ B^t & D \end{pmatrix}$. The matrices A and D are symmetric. Then $2f(x; y) = x^t Ax + y^t Dy + x^t By + y^t B^t x$ and $g(x; y) = x'^t x + y'^t y = x^t x' + y^t y'$. We have

$$\{f, g\}(x; y) = x^t(-Ay' + Bx') + y^t(-B^t y' + Dx').$$

Then $\{f, g\}(x; y) = X^t X'' = (X'')^t X$, where $X'' = M'' X'$, with $M'' = \begin{pmatrix} B & -A \\ D & -B^t \end{pmatrix}$.

We have $M = -2JM' = \begin{pmatrix} -B^t & -D \\ A & B \end{pmatrix}$ and $-M^t = \begin{pmatrix} B & -A \\ D & -B^t \end{pmatrix}$. We have finally $M'' = -M^t$. This ends the proof of claim *(i)*.

Claim *(ii)* follows from the equality

$$-J\begin{pmatrix} B & -A \\ D & -B^t \end{pmatrix} J = \begin{pmatrix} -B^t & -D \\ A & B \end{pmatrix}$$

$$-JM''J = M.$$ □

Lemma 3.2 *Let $\phi : sp(V) \to S^2V^*$ be the isomorphism of complex vector spaces defined by $\psi = id_{V^*} \otimes \frac{1}{2}\flat$. If $sp(V)$ is endowed with its Lie algebra structure and if S^2V^* is endowed with its Poisson algebra structure, then ϕ is an isomorphism of Lie algebras.*

Proof. We can define a Poisson action of S^2V^* on $V^* \otimes V^*$: $\{f, g \otimes h\} = \{f, g\} \otimes h + g \otimes \{f, h\}$.

Using the preceding lemma we see that the action of $sp(V)$ on $V^* \otimes V$ defined above corresponds to the Poisson action of S^2V^* on $V^* \otimes V^*$ modulo the isomorphisms $\phi : sp(V) \to S^2V^*$ and $id_{V^*} \otimes \frac{1}{2}\flat : V^* \otimes V \to V^* \otimes V^*$. (In claim *(ii)* of the preceding lemma, we can replace the isomorphism $\flat$ by the isomorphism $\frac{1}{2}\flat$.) The isomorphism $id_{V^*} \otimes \frac{1}{2}\flat : V^* \otimes V \to V^* \otimes V^*$ induces the isomorphism ϕ and the action of $u \in sp(V)$ on $sp(V)$ is $v \mapsto [u, v]$. This ends the proof of the lemma. □

We remark that the above lemma is nothing else but the formalization in our context of the well-known isomorphism between the Lie algebra of Hamiltonian linear vector fields and the Poisson algebra of quadratic Hamiltonian functions.

Now we can state the main result of this section.

Let $u \in sp(V)$. Using the action of u on V^* defined above we get an action on $\mathbf{C}[V]$. We denote by r this action $f \mapsto u.f$.

Let $\mathcal{G} \subset sp(V)$ be a Lie subalgebra.

We recall that $f \in \mathbf{C}[V]$ is an *invariant* of $\mathcal{G}$ if $u.f = 0$ for every $u \in \mathcal{G}$.

Theorem 3.6 *Let V be a symplectic complex space. We set $\dim_{\mathbf{C}} V = 2n$. Let $\mathcal{G} \subset sp(V)$ be a Lie subalgebra. Let $\alpha = (\alpha_1, \ldots, \alpha_n)$ be a finite involutive subset. We assume that the n elements $\alpha_1, \ldots, \alpha_n$ are algebraically independent and invariants of $\mathcal{G}$. Then the Lie algebra $\mathcal{G}$ is abelian.*

Proof. By Lemma 3.1 we see that $u.f = 0$ is equivalent to $\{\phi(u), f\} = 0$. We denote by A the subalgebra generated by α. This algebra is involutive, $A^{\perp}$ is involutive by Corollary 3.2 and $\phi(\mathcal{G}) \subset A^{\perp}$. As $\phi(\mathcal{G})$ is a Poisson algebra isomorphic to the Lie algebra $\mathcal{G}$, the result follows. $\square$

Now we are going to finish the proof of Corollary 3.2.

Let $U \subset \mathbf{C}^n$ be a connected open subset. We denote by $\mathcal{O}(U)$ the $\mathbf{C}$-algebra of holomorphic functions on U. We denote by $\mathcal{M}(U)$ the field of meromorphic functions on U, i.e., the fraction field of $\mathcal{O}(U)$. Let $f_1, \ldots, f_m \in \mathcal{M}(U)$. In this context, we say that they are *functionally dependent* if there exists a non-trivial relation $\sum_{i=1}^{m} g_i df_i = 0$, with $g_1, \ldots, g_m \in \mathcal{M}(U)$ $(g_1, \ldots, g_m \neq 0)$, that is if the meromorphic differential forms $df_1, \ldots, df_m$ are linearly dependent over the field $\mathcal{M}(U)$.

It is easy to prove the following result.

Lemma 3.3 *Let $U \subset \mathbf{C}^n$ be a connected open subset. Let $f_1, \ldots, f_m \in \mathcal{M}(U)$. The following conditions are equivalent:*

(i) *$f_1, \ldots, f_m$ are functionally independent;*

(ii) *there exists an open connected dense subset $W \subset U$ such that $f_1, \ldots, f_m \in \mathcal{O}(W)$ and $\mathrm{rank}_{\mathbf{C}}(df_1(x), \ldots, df_m(x)) = m$ for every $x \in W$;*

(iii) *there exists a point $x \in U$ such that $f_1, \ldots, f_m$ are holomorphic at x and such that $\mathrm{rank}_{\mathbf{C}}(df_1(x), \ldots, df_m(x)) = m$.*

Proposition 3.5 *In $\mathbf{C}(V) \approx \mathbf{C}(x_1, \ldots, x_n)$ functional (over some open set U of V) and algebraic dependence (over $\mathbf{C}$) are equivalent.*

This well-known result is proved (for instance) in [8], and it is also used in Ziglin's original paper [115]. We give the proof for completeness.

Proof. For proving the above proposition, we recall the following classical result ([8], Proposition 1.15).

Proposition 3.6 *Let $L \subset K$ be a field extension of 0-characteristic fields. Then any derivation on L extends to a derivation on K.*

Let $f_1, \ldots, f_m \in \mathbf{C}(V)$. If $P(f_1, \ldots, f_m) = 0$ for some polynomial

$$P \in \mathbf{C}[Z_1, \ldots, Z_m],$$

then $dP(f_1, \ldots, f_m) = \sum \frac{\partial}{\partial Z_i} P(f_1, \ldots, f_m) df_i = 0$, therefore $f_1, \ldots, f_m$ are functionally dependent.

Conversely, if the f_i's are algebraically independent then

$$\mathbf{C}(f) = \mathbf{C}(f_1, \ldots, f_m)$$

is a transcendental extension of degree m of $\mathbf{C}$ and the differential operators $\partial/\partial f_i$ are well defined on $\mathbf{C}(f)$. We have $\frac{\partial}{\partial f_i} f_j = \delta_{ij}$. We have a field extension $\mathbf{C}(f) \subset \mathbf{C}(V)$, therefore the differential operators $\partial/\partial f_i$ extend to derivations D_i on $\mathbf{C}(V)$. We have clearly $D_i f_j = \delta_{ij}$. We define germs of vector fields $X_i = \sum_{j=1,\ldots,n} D_i(x_j) \partial/\partial x_j$ ($i = 1, \ldots, m$). For every $g \in \mathbf{C}(V)$ we have $(dg, X_i) = D_i g$. Let $g_1, \ldots, g_m \in \mathbf{C}(V)$ be such that $\sum_{i=1,\ldots,m} g_i df_i = 0$. If we contract this relation with the vector field X_k, we get $g_k = 0$. Therefore the f_i's are functionally independent. $\qquad\square$

Proposition 3.7 *Let V be a symplectic complex space of dimension $2n$. Let $\mathbf{C}(V)$ be the field of rational functions on V. Let $f_1, \ldots, f_{n+1} \in \mathbf{C}(V) \approx \mathbf{C}(x_1, \ldots, x_{2n})$ in involution. Then the functions $f_1, \ldots, f_n, f_{n+1}$ are functionally dependent over an open domain $U \subset V$.*

This proposition is well known in the real (differentiable) case. In the complex case the same proof works well. We shall give the proof for completeness.

Proof. We can interpret $\flat$ as an isomorphism between the holomorphic fiber bundles TV (tangent bundle) and T^*V (cotangent bundle). We denote by $\natural$ the inverse isomorphism.

If we assume that the functions $f_1, \ldots, f_n, f_{n+1}$ are functionally independent, then they are regular and $\mathrm{rank}_{\mathbf{C}} (df_1, \ldots, df_{n+1}) = n+1$ on a dense open domain U.

The $\mathbf{C}$-linear forms $df_1(x), \ldots, df_{n+1}(x) \in T_x V^*$ are linearly independent for every $x \in U$. Let $x_0 \in U$. We set $f_i(x_0) = c_i \in \mathbf{C}$. The subset $\Sigma = \{f_1 = c_1, \ldots, f_{n+1} = c_{n+1}\} \subset U$ is an analytic (smooth) submanifold of complex dimension $n - 1$. The vector fields $Y_i = \natural df_i$ ($i = 1, \ldots, n+1$) are tangent to Σ ($df_i(Y_j) = \{f_i, f_j\} = 0$) and linearly independent over the complex field at each point of V (the linear map $\natural$ induces an isomorphism between $T_x V^*$ and $T_x V$). This implies $\dim \Sigma \geq n + 1$ and we get a contradiction. $\qquad\square$

Corollary 3.3 *Let V be a symplectic complex space. We set $\dim_{\mathbf{C}} V = 2n$. Let $f_1, \ldots, f_{n+1} \in \mathbf{C}(V) \approx \mathbf{C}(x_1, \ldots, x_{2n})$ in involution. Then*

(i) $f_1, \ldots, f_{n+1}$ are algebraically dependent,

(ii) if, moreover $f_1, \ldots, f_n$ are algebraically independent, then f_{n+1} is algebraic over the $\mathbf{C}$-algebra generated by $f_1, \ldots, f_n$.

Proof. The functions $f_1, \ldots, f_{n+1}$ are functionally dependent (on some open subset), therefore they are algebraically dependent (over the complex field $\mathbf{C}$).

If the functions $f_1, \ldots, f_n$ are algebraically independent, then we get a relation

$$P(f_1, \ldots, f_{n+1}) = A_m f_{n+1}^m + \cdots + A_0 = 0,$$

where $P \in \mathbf{C}[F_1, \ldots, F_{n+1}]$ and $A_0, \ldots, A_m \in \mathbf{C}[F_1, \ldots, F_n] \approx \mathbf{C}[f_1, \ldots, f_n]$. The last isomorphism is a consequence of the algebraic independence of $f_1, \ldots, f_n$, with $m > 0$.   □

With this result we can end the proof of Corollary 3.2 and therefore the proof of Theorem 3.6.

Chapter 4

Non-integrability Theorems

After the long preliminary work of Chapters 2 and 3, we now give the central theoretical results of this book. They will be used in a systematic way in the rest of this book.

In Section 4.1 we explain the process of reduction from the variational equations (VE) to the normal variational equations (NVE) within the framework of the differential Galois theory. It is important to recall that to some extent it can be considered as a formalization of the Poincaré theorem as explained in Chapter 3. This section contains an elementary but important new result that is essential in applications: if the identity component of the differential Galois group of the VE is abelian then the identity component of the differential Galois group of the NVE is also abelian.

The main theorems are given in Section 4.2. These theorems are to some extent variations, in different situations, of our central result: the non-abelian character of the identity component of the differential Galois group of the VE (or the NVE) is an obstruction to integrability in Liouville sense. The proof is based on an *infinitesimal* method: we analyze the structure of the Lie algebra of the Galois group. This approach is clearly allowed by the Galoisian formulation of Ziglin's theory: the differential Galois group is an algebraic group, and therefore a Lie group. In Ziglin's original formulation the monodromy group is discrete and it is impossible to use an infinitesimal method. We remark that the differential Galois group contains the monodromy group and therefore the Zariski closure of this monodromy group. But, in the irregular (i.e., non-Fuchsian) case, the differential Galois group can be *larger* than this Zariski closure. We stress that, if we search for a non-integrability criterion, then the larger the differential Galois group, the better!

Section 4.3 is devoted to some first examples in order to illustrate the above theoretical results. We will apply our non-integrability result to the fam-

J. J. Morales Ruiz, *Differential Galois Theory and Non-Integrability of Hamiltonian*, Systems, Modern Birkhäuser Classics, DOI: 10.1007/978-3-0348-0723-4_4, © Springer Basel 1999

ily of two degrees of freedom Hamiltonian potentials

$$U(x_1, x_2) = \frac{1}{3}x_1^3 + \frac{1}{2}(a + bx_1)x_2^2, \, a \in C^*, \, b \in C,$$

and

$$U(x_1, x_2) = \frac{1}{2}x_1^n + \frac{1}{2}(ax_1^{n-4} + bx_1^{n-3} + cx_1^{n-2})x_2^2, \, n \in \mathbf{N}, \, n > 3.$$

The results of this chapter were obtained in a joint work of the author with J.P. Ramis ([77]).

4.1   Variational equations

4.1.1   Singular curves

Let $X := X_H$ be a holomorphic Hamiltonian system defined on an analytic complex symplectic manifold M of dimension $2n$ (the phase space) by a Hamiltonian function H. Before going to formal constructions, we make some comments about the essential underlying ideas. If $x = \phi(t)$ is a germ of an integral curve, but not an equilibrium point, then one can consider the corresponding connected complete complex phase curve $i(\Gamma)$ in the phase space. We denote by Γ the corresponding *abstract* Riemann surface. By an abuse of terminology we say that Γ is an integral curve. From now on by an integral curve we understand, in general, this abstract curve. On the integral curve Γ we can use the complex time t, which is defined up to an additive complex constant, as a local parameter (uniformizing coordinate). However it is essential to think of Γ as an abstract Riemann surface over which we can use other local parametrizations. The only distinctive fact of a temporal parametrization is that it allows us to express the Hamiltonian field in the simplest way: $X = d/dt$. Then, using a pull-back, we interpret X as a holomorphic vector field on Γ.

If necessary we will make a careful distinction between the abstract Riemann surface Γ and the phase curve $i(\Gamma) \subset M$ which is the image of Γ by an immersion i. This immersion induces a bijection $\Gamma \to i(\Gamma)$. Note that i is not, in general, an embedding.

It can happen that the complex time is a *global* parametrization of Γ. This is frequent in applications (cf., Example 2 below). More precisely we have an analytic covering $\psi : \mathbf{C} \to \Gamma, \, t \mapsto \psi(t)$. But it is important to notice that, in general, in such a situation it is an infinitely sheeted covering (i.e., the function ψ is *transcendental*). We will see later that in our theory it is not important to distinguish the curves up to a finite covering (and we will use this fact in the applications very often), but it will be strictly forbidden to replace a curve by one of its infinitely sheeted coverings. Therefore in the preceding situation it is important to distinguish carefully between the integral curve Γ and the complex time line $\mathbf{C}$.

In the following, the variational equation over Γ is locally a system of linear equations with holomorphic coefficients, or more abstractly a holomorphic symplectic connection ∇ over Γ. We get it by pull-back from the variational equation over the phase curve $i(\Gamma)$, which is classically associated to our Hamiltonian system.

The following step is to introduce the possibility of adding singular points in order to obtain a *meromorphic* symplectic connection on some extended Riemann surface $\overline{\Gamma}$. It seems natural to add the equilibrium points of X that belong to the closure in the phase space M of the phase curve $i(\Gamma)$, i.e., the possible limit points of the phase curve $i(\Gamma)$ when the time is made infinite. The problem is that the resultant extended set is not, in general, an analytic smooth curve. We will limit ourselves to the following case: we suppose that the set of equilibrium points in the closure of $i(\Gamma)$ is discrete (if not, we add only a discrete subset) and that the extended curve $\underline{\Gamma}$ is an *analytic* complex subset of dimension one of M. We allow *singularities* on $\underline{\Gamma}$, and in general, they will be precisely the added equilibrium points. As usual in algebraic and analytic geometry, we can desingularize this curve and obtain a "good" Riemann surface $\overline{\Gamma}$. This Riemann surface is abstract and of course it is not contained in the phase space M. The holomorphic connection ∇ over Γ, which represents the VE, extends on a *meromorphic* connection over $\overline{\Gamma}$. The poles of this connection are "above" equilibrium points and correspond to branches of the curve $\underline{\Gamma}$ at the corresponding equilibrium point.

The reader not familiar with algebraic or analytic singular curves and their non-singular models can find some information in [53, 85, 108] (in particular [85] Theorem 4.1.11 is valid in our case if we replace the finite set of singular points by our discrete set and the compact analytic curve by our, in general, non-compact curve).

In some problems it is interesting to add *points at infinity* to Γ or to $\overline{\Gamma}$. We now add to the symplectic manifold (M, ω) a hypersurface at infinity M_∞: $M' = M \cup M_\infty$. We suppose that M' is a complex manifold and that ω admits a *meromorphic* extension over M'. Then it is natural to add to the curve $\underline{\Gamma}$ the points of M_∞ that belong to the closure of this curve in the extended phase space M'. The resultant extended set is not, in general, an analytic smooth curve. As before, we limit ourselves to the following case: we suppose that the set of points at infinity in the closure of $\underline{\Gamma}$ is discrete (if not, we add only a discrete subset) and that the extended curve $\underline{\Gamma}'$ is an *analytic* complex subset of dimension one of M'. Then, as before, we can desingularize this curve and obtain a Riemann surface $\overline{\Gamma}'$. The meromorphic connection ∇ over $\overline{\Gamma}$ which represents the VE extends on a meromorphic connection over $\bar{\Gamma}$. The poles of this connection are above equilibrium points or above points at infinity.

After these preliminaries, we start with the formal definitions. Let $x = \phi(t)$ (where t is a complex parameter, not necessarily the time) be a germ of regular holomorphic parametrized curve in the phase space, i.e., ϕ is given in local coordinates by a convergent Taylor expansion in a neighborhood of $t = 0$ with $\phi'(0) \neq 0$. We have locally a homeomorphism between an open disk centered in $t = 0$ and its image by ϕ. We consider two such elements ϕ_1, ϕ_2 as equivalent if there exists a germ of holomorphic function ρ, at the origin such that $\rho(0) = 0$, $\rho'(0) \neq 0$ and $\phi_2 = \phi_1 \circ \rho$ (change of parametrization). We denote by $\mathcal{C}$ the set of germs of curves over the phase space up to the above equivalence.

It is possible to endow the set $\mathcal{C}$ with a natural topology. If a germ ϕ belongs to $\mathcal{C}$, and is defined by a holomorphic function $\phi(t)$ for t varying in an open disk $U \subset \mathbf{C}$, then for any point t_0 in U, we define $\phi_{t_0} \in \mathcal{C}$ as $\phi_{t_0}(t) := \phi(t + t_0)$. The sets $U(\phi) := \{\phi_{t_0}\}$ are a basis for the open sets in $\mathcal{C}$.

Given a germ $\phi \in \mathcal{C}$, it defines the abstract Riemann surface Γ given by its connected component $i(\Gamma)$. For more details see [53], Chapter 7 (in this reference the analysis is made in the context of plane curves, but it remains clearly valid without changes in our situation) or the classical H. Weyl monograph ([108], p. 61).

So, if we have the germ of an integral curve that passes by a point x_0, $x = \phi_{x_0}(t)$ with the initial condition $\phi_{x_0}(0) = x_0$, then the Riemann surface which it defines is precisely Γ. If we identify Γ with the corresponding (connected) phase curve in the phase space $i(\Gamma)$, we get the Hamiltonian field X over Γ (d/dt in the temporal parametrization). More precisely this Hamiltonian field is the pull-back of the Hamiltonian field over M by the immersion $i : \Gamma \to M$. At the points of the set $\bar{\Gamma} - \Gamma$, the vector field X is by definition zero (as they correspond to the equilibrium points belonging to the closure of $i(\Gamma)$ in the phase space M).

Example. We illustrate the above considerations with an example that is important in the applications. Let a one degree of freedom system be defined by the following analytical Hamiltonian over an open set of $\mathbf{C}^2$

$$H(x, y) = \frac{1}{2}y^2 + \frac{1}{2}\varphi(x).$$

If we consider the energy level zero, we obtain an analytic subset defined by the equation $P(x, y) = y^2 + \varphi(x) = 0$. We assume that this set C is connected (if not, we select one of its connected components). Then C is an analytic curve. Its singular points are exactly the equilibrium points $E := \{(0, x) : \varphi(x) = \varphi'(x) = 0\}$. So, the curve $C = \underline{\Gamma}$ is equal to the disjoint union of Γ or more precisely $i(\Gamma)$ and E. And we obtain $\bar{\Gamma}$ as the corresponding non-singular model of $\underline{\Gamma}$.

In order to perform some explicit computations, we will analyze some simple particular cases:

1) Let $\varphi(x) = \frac{2}{3}x^3$. The curve C is now

$$P(x,y) = y^2 + \frac{2}{3}x^3.$$

And

$$C = \underline{\Gamma} = \Gamma \cup \{(0,0)\}, E = \{(0,0)\}.$$

We note that Γ admits a temporal parametrization

$$(x,y) = (-6t^{-2}, 12t^{-3}).$$

We can desingularize the point $(0,0)$, as usual, by using Puiseaux series ([53], Chapter 6), indeed we obtain only one branch $(x,y) = (-6\tilde{t}^2, 12\tilde{t}^3)$. We write $\tilde{t}$ instead of t because this parameter is not the time ($\tilde{t} = 1/t$), and this is so because the temporal parametrization of Γ is rational.

Now we can compute the Hamiltonian vector field X on $\overline{\Gamma}$. As $X = yd/dx$, on Γ, $X = d/dt$, if we use the temporal parametrization, and

$$\frac{y(\tilde{t})}{x'(\tilde{t})} \frac{d}{d\tilde{t}} = -\tilde{t}^2 \frac{d}{d\tilde{t}},$$

at the singular point s in $\overline{\Gamma} - \Gamma$.

2) Let $\varphi(x) = x^2(1-x)$. The curve $\underline{\Gamma}$ contains a homoclinic orbit $i(\Gamma)$, and the origin as an equilibrium point. We can parametrize *globally* the Riemann surface Γ using the time parametrization

$$(x(t), y(t)) = \left(\frac{2}{1+\cosh t}, -\frac{2\sinh t}{(1+\cosh t)^2}\right).$$

We remark that, despite its *transcendental appearence* (i.e., the above equations), the Riemann surface Γ is *algebraic*: it is only a problem related to the selected parametrization. Our global time parametrization is not one to one: it is an infinitely sheeted covering. Now applying the Puiseux algorithm we obtain immediately two branches and hence two points s_1, s_2 belonging to $\overline{\Gamma} - \Gamma$ above the origin in $\underline{\Gamma}$,

$$(x,y) = (\tilde{t}, \tilde{t} + h.o.t.),$$

$$(x,y) = (\tilde{t}, -\tilde{t} + h.o.t.).$$

We can express the field X as $(\tilde{t} + \cdots)d/d\tilde{t}$ or $(-\tilde{t} + \cdots)d/d\tilde{t}$, respectively.

3) We observe that in Example 1) the equilibrium point is degenerated and the field X has a zero of multiplicity two in $\overline{\Gamma}$ at the corresponding point above it. However, in Example 2) the equilibrium point is non-degenerated and the field X has a simple zero at the two corresponding points of $\overline{\Gamma}$. This is not casual. In fact, let $\varphi(x) = x^n + O(x^{n+1})$ (with $n \leq 2$) be the expansion of φ at the origin, then, by a simple inspection of the Newton polygon we get the following facts. If n is odd, we get only one point belonging to $\overline{\Gamma} - \Gamma$, $(x, y) = (\tilde{t}^2, \tilde{t}^n + h.o.t.)$, and the field X has a zero of order $n - 1$ at this point. If n is even, we get two points above $(0, 0)$ in $\overline{\Gamma} - \Gamma$, $(x, y) = (\tilde{t}, \tilde{t}^{n/2})$, $(x, y) = (\tilde{t}, -\tilde{t}^{n/2})$ and the field X has a zero of order $n/2$ at each one. Of course, the above results do not depend upon the parametrization. This simple fact is fundamental, as will become clear in the applications.

4.1.2 Meromorphic connection associated with the variational equation

Once we have defined Γ and the derivation X, we define the holomorphic connection associated to the variational equation (VE) over Γ. More generally, if we add some equilibrium points (respectively some equilibrium points and some points at infinity), we define the *meromorphic* connection associated to the variational equation (VE) over $\overline{\Gamma}$ (respectively $\overline{\Gamma}'$).

Let T_Γ be the restriction to Γ or more precisely to $i(\Gamma)$ of the tangent bundle TM to the phase space M. It is a symplectic holomorphic vector bundle. More formally the fiber bundle T_Γ is the pull-back of T_M by the immersion $i : \Gamma \to M$.

The holomorphic connection which defines the variational equation along Γ comes by pull-back from the restriction to $i(\Gamma)$ of the Lie derivative with respect to the field X

$$\nabla v := L_X Y_{|\Gamma},$$

where Y is any holomorphic vector field extension of the section v of the bundle $T_{i(\Gamma)}$.

The fact that the connection ∇ is symplectic follows from the definition of a Hamiltonian vector field: the symplectic form is preserved by the flow.

Using the local time parametrization $z = z(t)$ on Γ and local canonical coordinates $z = (x_1, \ldots, x_n, y_1, \ldots, y_n) := (z_1, \ldots, z_{2n})$ on M, we have the usual definition of the variational equation (VE) along the phase curve.

We write the Hamiltonian system

$$\dot{z} = J \frac{\partial H}{\partial z},$$

where $\frac{\partial H}{\partial z}$ is the Jacobian matrix of H. Then we can express our connection ∇ in a holomorphic frame $(e_1, \ldots, e_{2n})$, and we get a differential system. Choosing the symplectic frame $\frac{\partial}{\partial z_i}$ associated to the canonical coordinates $(z_1, \ldots, z_{2n})$, we get the differential system

$$\frac{d\xi}{dt} = A(t)\xi,$$

where

$$A(t) = \frac{\partial X}{\partial z}(z(t)) = J\,\mathrm{Hess}H(z(t)) = JS,$$

$S = \mathrm{Hess}H(z(t))$ being the Hessian matrix of the Hamiltonian function H (this is a direct consequence from the expression of the Lie bracket $[\sum X_i \partial/\partial z_i, \partial z_j]$). Hence we obtain the differential system that defines the VE in its usual form. It is clearly a linear Hamiltonian system.

Now we add to $i(\Gamma)$ a discrete set of *equilibrium* points, where by definition the field X_H vanishes. We suppose that $\underline{\Gamma}$ is the closure, in the phase space M, of $i(\Gamma)$ and that $\underline{\Gamma}$ is, in general, an analytic curve in M which is singular at the equilibrium points. We denote by $\bar{\Gamma} \to \underline{\Gamma}$ a desingularization of the curve $\underline{\Gamma}$. We consider Γ as an open subset of $\bar{\Gamma}$. By restricting the tangent bundle TM to $\underline{\Gamma}$, we get an holomorphic bundle $T_{\underline{\Gamma}}$ over $\underline{\Gamma}$, and by pull-back a holomorphic bundle $T_{\bar{\Gamma}}$ over $\bar{\Gamma}$.

If the point a above is an equilibrium point, we cannot use the temporal parametrization on $\bar{\Gamma}$. Then we choose a uniformizing variable u at $a \in \bar{\Gamma}$ $(u(a) = 0)$, and we write the immersion $i: u \mapsto i(u)$. We get $X(i(u)) = f(u)d/du$ $(u \neq 0)$ with $f(0) = 0$ (by definition the Hamiltonian field X vanishes at $i(0)$). Using u as a local coordinate on $i(\Gamma)$ we can write $X = d/dt = f(u)d/du$. Then the pull-back by i of the VE in a punctured neighborhood of a is

$$\frac{d\xi}{du} = \frac{1}{f(u)}\frac{\partial X}{\partial x}(i(u)).$$

This system is a *holomorphic* differential system over a punctured neighborhood of a in $\bar{\Gamma}$. It can clearly be interpreted as a *meromorphic* differential system over a neighborhood of a.

Such local constructions over $\bar{\Gamma}$ glue together and we get a symplectic meromorphic connection over $\bar{\Gamma}$. It defines the VE over $\bar{\Gamma}$.

When some points at infinity are added, a similar construction over $\bar{\Gamma}'$ can be performed. The only difference is that the Hamiltonian field $X = X_H$ can have a pole at one of these points at infinity due to possible singularities at infinity of the symplectic form. Then the function $f(u)$ is now in general *meromorphic*.

4.1.3 Reduction to normal variational equations

The problem of reducing a linear system of equations goes back to D'Alembert's reduction of the order of a linear differential equation by means of a known particular solution.

In the Hamiltonian case, as we shall see, the underlying mechanism that explains reduction is the existence of invariant unidimensional horizontal sections (of (V, ∇) or (V^*, ∇^*)) which are in involution.

All the bundles and connections considered in this section are meromorphic. In the process of reduction we might need to add some new singular points of the reduced connection. We consider these new singularities as singular points of the initial VE, i.e., as points of $\overline{\Gamma} - \Gamma$. With this in mind, all the bundles and connections will be defined over the same fixed connected Riemann surface $\overline{\Gamma}'$ or $\overline{\Gamma}$ with the same "singular set". Also, as usual, we will identify a bundle with its sheaf of sections.

Let V be a symplectic vector bundle of rank $2n$. Locally we can define a symplectic form Ω which defines the symplectic structure of V. If $v_1, \ldots, v_k$ are *global* meromorphic sections of V linearly independent over Γ ($v_1 \wedge \cdots \wedge v_k \in \bigwedge^k V$ is different from zero on Γ) and in involution (i.e., $\Omega(v_i, v_j) = 0$, $i, j = 1, \ldots, k$), then we can obtain some subbundles of V as follows. But before this we remark that, by definition, the sections $v_1, \ldots, v_k$ have their coefficients in the field of meromorphic functions over $\overline{\Gamma}$ or $\overline{\Gamma}'$, being holomorphic over Γ.

Let F be the rank k (meromorphic) subbundle of V generated by $v_1, \ldots, v_k$. We get $F^\perp$ ($\perp$ with respect to the symplectic structure) as a sub-bundle of V. We have clearly $F \subset F^\perp$. The normal bundle $N := F^\perp / F$ is a rank $2(n-k)$ symplectic bundle which admits locally the symplectic form Ω_N defined by the projection of Ω. It is easy to see that this form is well defined and non-degenerated over Γ. See also [8, 65], where meromorphic vector bundles do not appear but the constructions are similar.

Using Lemma 4.2 below, we can suppose that the form Ω_N is globally defined and that it is meromorphic over $\overline{\Gamma}$ and holomorphic and non-degenerated over Γ. We will implicitly make these hypotheses in what follows.

The following proposition (with the same notation as above) is essentially a consequence of Propositions 1.11 and 1.6 of [8] with small changes in the notation and taking into account the fact that we work here with meromorphic connections instead of holomorphic connections.

Proposition 4.1 *Let* (∇, V, Ω) *be a symplectic connection and* $v_1, \ldots, v_k$ *an involutive set of linearly independent global horizontal sections of* ∇. *Then, by restriction, we have the subconnections* $(\nabla_F = 0, F)$, $(\nabla_{F^\perp}, F^\perp)$ *and a symplectic connection,* (∇_N, N, Ω_N), *on the normal bundle.*

Proof. It is obvious that the bundle F is invariant by ∇. The invariance of $F^\perp$ by ∇ follows from the formula

$$\Omega(\nabla w, v) = X(\Omega(w,v)) - \Omega(w, \nabla v) - (\nabla \Omega)(w,v),$$

and from the fact that ∇ is a symplectic connection.

We define the connection ∇_N on $N := F^\perp/F$ from the action of ∇ on the representatives of the classes of N in $F^\perp$. Of course ∇_N is well defined ($\nabla_F = 0$) and it is a symplectic connection. $\qquad\square$

The connection ∇_N is called the normal (reduced) connection and the corresponding linear differential equation the normal equation.

We remark that although the proof of the above proposition is technically similar to those of the propositions in [8], the philosophy here is different. Here the connections ∇, ∇_N have the same singularities in $\overline{\Gamma} - \Gamma$; in particular, the differential fields of coefficients of the corresponding linear differential equations are the same (the meromorphic functions over $\overline{\Gamma}$).

Our objective now is to investigate the relation between the Galois group of the initial equation Gal ∇ and the Galois group of the normal equation Gal ∇_N. We will use two different methods. The first is based upon explicit classical computations. The results so obtained are sufficient for the applications. More precisely for these applications it is sufficient to know that if the identity component of Gal ∇ is abelian, then the identity component of Gal ∇_N is also *abelian*. The second method is based upon Tannakian arguments and a more precise relation between the two differential Galois groups is established. This relation is interesting by itself, although not necessary for the work being developed here.

Let (∇, V, Ω) be a symplectic connection. Then we have the musical isomorphism defined by Ω

$$\flat : (\nabla, V, \Omega) \longrightarrow (\nabla^*, V^*, \{,\}),$$

so the symplectic form section is transported to the Poisson bracket,

$$\{\alpha, \beta\} = \Omega(\natural(\alpha), \natural(\beta)),$$

$\natural := \flat^{-1}$. In some references it is said that V^* with the Poisson bracket is a Poisson vector bundle, see for instance [65].

Now, let $\alpha \in V^*$, $v \in V$ be two sections. Then $\alpha_0 := \alpha(p_0)$, $v_0 := v(p_0)$ are elements of the fibres at $p_0 \in \Gamma$, and by using Cauchy's existence theorem can be identified with elements belonging, respectively, to the vector spaces of germs of solutions at p_0: $E^* = \text{Sol}\,\nabla^*$, $E = \text{Sol}\,\nabla$.

Lemma 4.1 *Let* (V, ∇, Ω) *be a symplectic connection and let* α, $v := \flat^{-1}\alpha$, *global sections of, respectively, the bundles* V^* *and* V. *Then the following conditions are equivalent:*

  (i) α *is a (linear) first integral of the linear equation defined by* ∇.
  (ii) α *is a horizontal section of* ∇^* *(i.e., a solution of the adjoint differential equation).*
 (iii) v *is a horizontal section of* ∇.
 (iv) α_0 *is invariant by the Galois group* Gal ∇.
  (v) v_0 *is invariant by the Galois group* Gal ∇.

Proof. The equivalence between (i) and (ii) follows from

$$X(\langle \alpha, w \rangle) = \langle \nabla\alpha, w \rangle + \langle \alpha, \nabla w \rangle,$$

X being the holomorphic vector field on the Riemann surface $\overline{\Gamma}$, such that $\nabla := \nabla_X$. The equivalence between (ii) and (iii) follows from

$$0 = X(\Omega(v, v)) = X(\langle \alpha, v \rangle) = \langle \nabla\alpha, v \rangle + \langle \alpha, \nabla v \rangle.$$

Now if α is a horizontal section of ∇^*, $(\mathcal{M}_{\overline{\Gamma}}(1 + \alpha), \delta \oplus \nabla^*)$ is a rank one subconnection of $(\mathcal{M}_{\overline{\Gamma}} \oplus V^*, \delta \oplus \nabla^*)$. Then (as in the final example of Section 2.4) the corresponding complex construction by the fibre functor Sol, that is the complex line $\mathbf{C}(1 + \alpha_0)$, is (pointwise) invariant by the Galois group. So we get (iv). That (ii) is a necessary condition for (iv) is clear from the fact that α_0 is a local horizontal section at a point $p_0 \in \Gamma$ that, by assumption, can be extended to a global section α. From the unicity in Cauchy's theorem it is necessarily a horizontal section.

The equivalence between (iii) and (v) is obtained in a similar way: we only need to write V instead of V^* (another way of finishing the proof is to prove the equivalence between (iv) and (v) using the fact that the musical isomorphism $\flat_0$ between the vector spaces (E, Ω_0) and $(E^*, \{, \}_0)$ induces a bijection between the invariants of the Galois group in E and E^*). $\qquad\square$

Let $\alpha_1, \ldots, \alpha_k$ (α_i section of V^*) be an involutive set of (global) independent (i.e., they generate a rank k subbundle) first integrals of the symplectic connection (∇, V, Ω). By the above lemma we obtain an involutive set $v_1, \ldots, v_k$ of independent (global) horizontal sections of (∇, V, Ω). If as above, F is the rank k subbundle of horizontal sections generated by $v_1, \ldots, v_k$, we can construct the subbundles and connections (∇_F, F), $(\nabla_{F^\perp}, F^\perp)$ and $(\nabla_N, N = F^\perp/F, \Omega_N)$ (in general we will write simply ∇_N). We remark that it is easy to prove [8] that

$$F^\perp = \{w \in V : \langle \alpha_i, w \rangle = 0, i = 1, \ldots, k\}.$$

From the (meromorphic) triviality of the symplectic vector bundles (see Appendix B) and the properties of the symplectic bases (by Proposition 2.4, the global meromorphic sections are a symplectic vector space over the field of global meromorphic functions $K = \mathcal{M}(\overline{\Gamma})$ over $\overline{\Gamma}$) we have

Lemma 4.2 *There exists a global (meromorphic) symplectic canonical frame that contains the given linearly independent and involutive horizontal sections* $v_1, \ldots, v_k$.

Now we compute the normal equation in coordinates. Let JS be the matrix of ∇ in a canonical frame (S is a symmetric matrix). We define a symplectic change of variables using some new canonical frame which contains the given linearly independent and involutive horizontal sections $v_1, \ldots, v_k$.

We denote the matrix of the symplectic change of variables by

$$P = (D_1 \ D_2 \ C_1 \ C_2),$$

where $C_2 = (\xi_1^t, \ldots, \xi_k^t)$, and the $2n$-dimensional column vectors ξ_i, $i = 1, \ldots, k$, are the coordinates of v_i in the original canonical frame. Then we have [72]:

$$P^{-1} = \begin{pmatrix} -C_1^t J \\ -C_2^t J \\ D_1^t J \\ D_2^t J \end{pmatrix},$$

$$AP - \dot{P} = JSP - \dot{P} = (JSD_1 - \dot{D}_1 \quad JSD_2 - \dot{D}_2 \quad JSC_1 - \dot{C}_1 \quad 0),$$

since C_2 is a fundamental matrix solution of the original linear equation. Hence the matrix of the transformed equation is

$$P[JS] := P^{-1}(JSP - \dot{P}) = \begin{pmatrix} C_1^t(SD_1 + J\dot{D}_1) \ E & C_1^t(SC_1 + J\dot{C}_1) & 0 \\ C_2^t(SD + J\dot{D}) & C_2^t(SC_1 + J\dot{C}_1) & 0 \\ -D_1^t(SD_1 + J\dot{D}_1) \ F & -D_1^t(SC_1 + J\dot{C}_1) & 0 \\ -D_2^t(SD + J\dot{D}) & G & 0 \end{pmatrix},$$

where $D = (D_1 \ D_2)$ and E, F, G are given matrices. The matrix $P[JS]$ is necessarily infinitesimally symplectic, i.e., of the form JS_1 with S_1 symmetric (see, for instance, [72], pp. 36, 37). Hence $C_2^t(SD + J\dot{D})$ and $C_2^t(SC_1 + J\dot{C}_1)$ are zero, $E = -G^t$, $M = F^t$, $C_1^t(SD_1 + J\dot{D}_1) = (C_1^t S - \dot{C}_1^t J)D_1$ and H, $C_1^t(SC_1 + J\dot{C}_1)$, $-D_1^t(SD_1 + J\dot{D}_1)$ are symmetric.

Reordering the new canonical frame, the matrix of the connection ∇ becomes

$$P[JS] = \begin{pmatrix} C_1^t(SD_1 + J\dot{D}_1) & C_1^t(SC_1 + J\dot{C}_1) & G & 0 \\ -D_1^t(SD_1 + J\dot{D}_1) & (-D_1^t S - \dot{D}_1^t J)C_1 & F & 0 \\ 0 & 0 & 0 & 0 \\ F^t & -G^t & H & 0 \end{pmatrix}.$$

Then the transformed differential equation $\dot{\eta} = P[JS]\eta$ in the variable $\eta = (\alpha, \beta, \gamma, \delta)$ is

$$\begin{array}{rcl}
\dot{\alpha} & = & C_1^t(SD_1 + J\dot{D}_1)\alpha + C_1^t(SC_1 + J\dot{C}_1)\beta + G\gamma, \\
\dot{\beta} & = & -D_1^t(SD_1 + J\dot{D}_1)\alpha + (-D_1^tS - \dot{D}_1^tJ)C_1\beta + F\gamma, \\
\dot{\gamma} & = & 0, \\
\dot{\delta} & = & F^t\alpha - G^t\beta + H\gamma.
\end{array}$$

The matrix of the normal equation is

$$\begin{pmatrix} C_1^t(SD_1 + J\dot{D}_1) & C_1^t(SC_1 + J\dot{C}_1) \\ -D_1^t(SD_1 + J\dot{D}_1) & (-D_1^tS - \dot{D}_1^tJ)C_1 \end{pmatrix}.$$

Then we get the Picard-Vessiot extension L/K of ∇ from two successive extensions

$$K \subset L_N \subset L,$$

where L_N/K is the Picard-Vessiot extension of the normal equation (i.e.,of ∇_N) and L/L_N is an extension composed by two successive extensions L/L_1 and L_1/L_N by integrals.

Now, using the Galois correspondence, we get

$$\mathrm{Gal}(L_N/K) = \mathrm{Gal}(L/K)/\mathrm{Gal}(L/L_N).$$

It is well known that extensions by integrals are normal purely transcendental extensions. Their Galois groups are additive abelian groups isomorphic to some $G_a^r := (\mathbf{C}^r, +)$ (see Section 2.2). Therefore $\mathrm{Gal}(L/L_N)$ is Zariski connected and we get the inclusion of $\mathrm{Gal}(L/L_N)$ in the identity component $\mathrm{Gal}\,\nabla_0$ ($\mathrm{Gal}(L/L_N)$ is the group H that will appear below in the Tannakian reduction: see the next subsection).

Finally, for $n = k$ we can solve the initial system by quadratures. This is proved in [72]. Our computations on the matrix JS are a generalization of the computations in this reference (we in fact kept the same notations).

Proposition 4.2 *Let $\alpha_1, \ldots, \alpha_k \in V^*$ be an involutive set of independent (global) first integrals of (∇, V, Ω). Let ∇_N be the normal connection defined by the above set. Then we have*

(i) *The linear differential equation corresponding to the connection (∇, V, Ω) is integrable if, and only if, the normal equation corresponding to (∇_N, N, Ω_N) is integrable.*

(ii) *If the identity component of $\mathrm{Gal}\,\nabla$ is abelian then the identity component of $\mathrm{Gal}\,\nabla_N$ is also abelian.*

Proof. Using the preceding results and the differential Galois correspondence, the equivalence (i) follows from the general group theory, i.e., any quotient group of a solvable group is solvable, and conversely, if a normal subgroup and the corresponding quotient group are solvable then the original group is also solvable. Claim (ii) is evident. □

Claim (ii) in the preceding proposition is *essential* for applications.

If now we have the variational equation (VE) over an integral curve Γ of a meromorphic Hamiltonian system X_H, and if $f_1 = H, f_2, \ldots, f_k$ is an involutive set of independent meromorphic first integrals of X_H, then we get an involutive set $\alpha_1 = dH, \alpha_2 = df_2, \ldots, \alpha_k = df_k$ of independent first integrals of the (VE) (see Section 4.2) and we can apply the results of this section. The normal equation so obtained is then called the *normal variational equation* (NVE).

Example. As an example, we apply the above considerations to the variational equations along an integral curve Γ of a two degrees of freedom Hamiltonian system. We want to obtain the (NVE) from the (VE) when the Hamiltonian is a natural mechanical system, $H = T + U$, $T = \frac{1}{2}(y_1^2 + y_2^2)$, $U = U(x_1, x_2)$ (potential).

Then $\alpha = dH$ (over $\overline{\Gamma}$) is a linear first integral of the (VE), i.e., an element of V^*. We know that this is equivalent to the fact that $\dot{z} = (y_1, y_2, -U_1, -U_2) \in V$ (where we have used subindexes for the derivatives of the potential) is a known solution. There are several possibilities for the choice of the symplectic change P (which is defined by a symplectic frame admitting $\dot{z}$ as one of its elements). We can suppose that y_1 and y_2 are not identically zero (if this is not the case the NVE is obtained without any computation from the VE), then we can select a very simple solution

$$
P = \begin{pmatrix} 0 & 0 & 0 & y_1 \\ -\frac{y_2}{y_1} & 0 & 0 & y_2 \\ 0 & -\frac{1}{y_1} & 1 & -U_1 \\ \frac{U_2}{y_1} & 0 & -\frac{y_1}{y_2} & -U_2 \end{pmatrix}.
$$

Applying the formula obtained above, the matrix of the NVE is

$$
\begin{pmatrix} -U_1/y_1 & 1 + (y_1/y_2)^2 \\ y_2 U_{12}/y_1 & U_1/y_1 \end{pmatrix}.
$$

We observe that, as expected, it belongs to the symplectic Lie algebra $sp(1, K) = sl(2, K)$.

4.1.4   Reduction from the Tannakian point of view

We now give a new proof of Proposition 4.2 using Tannakian arguments. In fact we will get a slightly more precise result. This improvement even if it is not used later in this book, is important by itself.

We recall that a group G is *meta-abelian* if its derived group G' is abelian. In particular, a meta-abelian group is solvable.

The relation between the initial Galois group and the reduced Galois group is given by the following result

Proposition 4.3 *Let $\alpha_1, \ldots, \alpha_k \in V^*$ be an involutive set of independent (global) first integrals of (∇, V, Ω). Let ∇_N be the normal connection defined by means of the above set. Then*

$$(\mathrm{Gal}\,\nabla_N)_0 \approx \mathrm{Gal}\,\nabla_0/H,$$

where $\mathrm{Gal}\,\nabla_0$ and $\mathrm{Gal}\,\nabla_{N0}$ are respectively the identity components of the Galois groups of ∇ and ∇_N, and H is a closed normal meta-abelian subgroup of $\mathrm{Gal}\,\nabla_0$.

Proof. With the above notation, at the level of connections we have the natural morphisms

$$(V, \nabla) \hookleftarrow (F^\perp, \nabla_{F^\perp}) \to (N = F^\perp/F, \nabla_N),$$

(the first one is the inclusion and the second the projection) and the isomorphism

$$\flat : (\nabla, V, \Omega) \longrightarrow (\nabla^*, V^*, \{,\}).$$

Applying the fibre functor we get the corresponding morphisms and isomorphism

$$(E, \Omega) \hookleftarrow (F_0^\perp, \Omega_{F_0^\perp}) \to (N_0 = F_0^\perp/F_0, \Omega_{N_0})$$

$$\flat_0 : (E, \Omega) \longrightarrow (, V_0^*, \{,\}),$$

where $E = \mathrm{Sol}\,\nabla$, $F_0 = \mathrm{Sol}\,\nabla_F$, etc.... In order to simplify the notation, we write Ω and $\{,\}$ instead of Ω_0, $\{,\}_0$.

Let $F_0^*, \{,\}$ be the involutive subalgebra generated by $\alpha_1, \ldots, \alpha_k$. Then we obtain the morphisms

$$(E^*, \{,\}) \hookleftarrow (F_0^{*\perp}, \{,\}_{F_0^{*\perp}}) \to (N_0^* := F_0^{*\perp}/F_0^*, \{,\}_{N_0^*}),$$

where the orthogonality $\perp$ is now defined by the Poisson bracket.

We have the natural morphisms of algebraic groups

$$\mathrm{Gal}\ \nabla \to \mathrm{Gal}\ \nabla_{F^\perp} \to \mathrm{Gal}\ \nabla_N.$$

By composition, we get a surjective morphism $\phi : \mathrm{Gal}\ \nabla \to \mathrm{Gal}\ \nabla_N$.

We get also the corresponding morphisms of Lie algebra

$$\operatorname{Lie} \nabla \to \operatorname{Lie} \nabla_{F^\perp} \to \operatorname{Lie} \nabla_N$$

and the surjective morphism

$$\pi : \operatorname{Lie} \nabla \to \operatorname{Lie} \nabla_N.$$

We know from Section 3.4 that the Lie algebra $\operatorname{Lie} \nabla$ is isomorphic to a Lie subalgebra of $(S^2 E^*, \{,\})$ and that, modulo this isomorphism, the action of $\beta \in \operatorname{Lie} \nabla$ on E^* is given by

$$\{\beta,.\} : \delta \mapsto \{\beta, \delta\}.$$

Then it is easy to describe the natural morphisms

$$\operatorname{Lie} \nabla \to \operatorname{Lie} \nabla_{F^\perp} \to \operatorname{Lie} \nabla_N$$

by restriction and projection of $\{\beta,\}$ (we observe that $\{\beta, \alpha\} = 0$, for any $\beta \in \operatorname{Lie} \nabla$, $\alpha \in F_0^*$ or in a shorter way $\{\operatorname{Lie} \nabla, F_0^*\} = 0$), and $\operatorname{Lie} \nabla_N$ is considered also as a subalgebra of $(S^2 N^*, \{,\})$, modulo a musical isomorphism.

Using the morphism of Lie algebras

$$\pi : \operatorname{Lie} \nabla \to \operatorname{Lie} \nabla_N$$

we get an isomorphism

$$\operatorname{Lie} \nabla_N \approx \operatorname{Lie} \nabla / \operatorname{Ker} \pi.$$

We set $\ker \pi := \mathcal{H}$. Then $\beta \in \mathcal{H}$ if and only if $\{\beta, F_0^{*\perp}\} \subset F_0^*$. As $\{E^*, E^*\} \subset \mathbf{C}$, by the Jacobi identity

$$\{\{\beta, \delta\}, \alpha\} = \{\beta, \{\delta, \alpha\}\} - \{\delta, \{\beta, \alpha\}\},$$

(with $\beta \in \operatorname{Lie} \nabla$, $\delta \in E^*$, $\alpha \in F_0^*$), we get the inclusion $\{\operatorname{Lie} \nabla, E^*\} \subset F_0^{*\perp}$. Applying again the Jacobi identity in the form

$$\{\{\beta, \beta'\}, \delta\} = \{\beta, \{\beta', \delta\}\} - \{\beta', \{\beta, \delta\}\},$$

with $\beta, \beta' \in \mathcal{H}$ and $\delta \in E^*$, we note that this identity is simply the expression of the action of the commutator $[A, B]$ as $AB - BA$ in the usual linear representation; we obtain $\{\{\beta, \beta'\}, \delta\} \in F_0^*$. So, as the algebra $\mathcal{H}$ annihilates F_0^* we conclude that the derived algebra $\mathcal{H}'$ is abelian, i.e., $\mathcal{H}$ is meta-abelian and, in particular, solvable.

We set $H = \ker \phi$. The group H is an algebraic subgroup of $\operatorname{Gal} \nabla$. Using the results of the preceding subsection we can interpret H as a differential Galois group and we recall that this group is Zariski connected. Its Lie algebra is $\mathcal{H}$, therefore H is meta-abelian. $\qquad\square$

From the classical Picard-Vessiot theory we get

Corollary 4.1 *We have the following statements:*
  (i) *The linear differential equation corresponding to the connection*
      (∇, V, Ω) *is integrable if and only if the normal equation corresponding to*
      (∇_N, N, Ω_N) *is integrable.*
  (ii) *If the identity component of* Gal ∇ *is abelian then both the identity component of* Gal ∇_N *and the group* H *are abelian.*

Proof. (i) follows as in Proposition 4.2 (but using here Proposition 4.3). Claim (ii) is a direct consequence of the above proposition.   □

In the above corollary (i) expresses the meaning of the reduction: we can solve (in the Galois differential sense) the linear equation corresponding to ∇ when we know the solutions of the linear equation corresponding to ∇_N.

4.2   Main results

Let E be a complex vector space of dimension $m \geq 1$. As above we denote by $\mathbf{C}[E]$ the $\mathbf{C}$-algebra of polynomial functions on E, and by $\mathbf{C}(E)$ the field of rational functions on E, i.e., the quotient field of $\mathbf{C}[E]$.

Let $G \subset GL(E)$ be an algebraic subgroup. We define a left action of G on $\mathbf{C}[E]$ or $\mathbf{C}(E)$ by $(g, f) = g.f = f \circ g^{-1}$ $(g \in G, f \in \mathbf{C}(E))$. It corresponds to the usual action of G on the constructions over E. Let $\mathcal{G}$ be the Lie algebra of G. If $u \in \mathcal{G} \subset \mathrm{End}(E)$, we define its action on V^* by $-u^t$ and its action on $E^* \otimes E^*$ by $-u^t \otimes 1 - 1 \otimes u^t$. Its natural action $f \mapsto u \bullet f$ on $\mathbf{C}[E]$ (isomorphic to the symmetric tensor algebra $S^*(E^*)$) or on $\mathbf{C}(E)$ follows using evident formulas.

We define by $\mathbf{C}[E]^G$ (respectively $\mathbf{C}(E)^G$) the $\mathbf{C}$-algebra of G-invariant elements of $\mathbf{C}[E]$ (resp. $\mathbf{C}(E)$) (i.e., those $f \in \mathbf{C}[E]$ (respectively $\mathbf{C}(E)$) such that $g.f = f$, for all $g \in G$). If $f \in \mathbf{C}(V)^G$, then $u \bullet f = 0$ for all $u \in \mathcal{G}$.

We can clearly identify $\mathbf{C}$ with a subfield of $\mathbf{C}(E)^G$. As in [25, 8], for $r \geq 1$, we will say that an algebraic group G is r-Ziglin if trans $\deg_{\mathbf{C}} \mathbf{C}(E)^G \geq r$. We will say that an algebraic group G is r-involutive Ziglin if there exists r algebraically independent elements $f_1, \ldots, f_r \in \mathbf{C}(E)^G$ in involution.

We can now state our main results.

In order to facilitate the main applications to non-academic problems, we will give three versions of this main result and add some corollaries for local situations. Each statement will generalize the preceding.

Let Γ be the abstract Riemann surface defined by a non-equilibrium connected integral curve $i(\Gamma)$ of an analytic Hamiltonian system X_H with n degrees of freedom on a symplectic complex manifold M.

Theorem 4.1 *If there are n meromorphic first integrals of X_H that are in involution and independent over a neighborhood of the curve $i(\Gamma)$, not necessarily on Γ itself, then the Galois group of the VE over Γ is n-involutive Ziglin. This Galois group is the Zariski closure of the monodromy group. Furthermore, the identity component of the Galois group of the VE over Γ is abelian.*

We remark that in this theorem we take as coefficient field of the VE the meromorphic functions over Γ.

The following result is a consequence of Ziglin's theorem ([114] Theorem 2, Remark 1). As an exercise, we give here a different proof (in fact Ziglin's result is stronger: he does not need the involutivity hypothesis).

Corollary 4.2 *We suppose that there are n meromorphic first integrals of X_H that are in involution and independent over a neighborhood of the curve $i(\Gamma)$ (not necessarily on Γ itself). Let g, g' be non-resonant elements of the monodromy group, then g must commute with g'.*

Proof. Let G' be the Zariski closure in the Galois group G of the subgroup generated by g and g'. Let H (respectively H') be the Zariski closure of the subgroup generated by g (respectively g'). Because g and g' are non-resonant, the groups H and H' are maximal tori in the symplectic group, i.e., they are conjugated to the multiplicative group of dimension n

$$T := \{\operatorname{diag}(\lambda_1, \ldots, \lambda_n, \lambda_1^{-1}, \ldots, \lambda_n^{-1}, \lambda_i \in \mathbf{C}^*, i = 1, \ldots, n\}.$$

(see [8], Proposition 2.13)

Therefore they are Zariski-connected. It follows that $G' \subset G^0$. Applying Theorem 4.1 we see that G' is abelian, therefore g and g' commute. □

We now add to the curve $i(\Gamma)$ a discrete set of equilibrium points. We get a singular curve $\underline{\Gamma} \subset M$. Let $\bar{\Gamma}$ be a non-singular model of $\underline{\Gamma}$.

Theorem 4.2 *If there are n first integrals of X_H which are meromorphic and in involution over a neighborhood of the curve $\underline{\Gamma}$ and independent in a neighborhood of Γ (not necessarily on Γ itself), then the Galois group of the meromorphic VE over $\bar{\Gamma}$ is n-involutive Ziglin. Furthermore, the identity component of the Galois group of the VE over $\bar{\Gamma}$ is abelian.*

We add now to the symplectic manifold (M, ω) a hypersurface at infinity M_∞ ($M' = M \cup M_\infty$) and to the curve $i(\Gamma)$ a discrete set of equilibrium points and a discrete set of points at infinity. We get a singular curve $\underline{\Gamma}' \subset M'$, where, as before, we suppose that ω admits a meromorphic extension over M'. Let $\bar{\Gamma}'$ be a non-singular model of $\underline{\Gamma}'$.

Theorem 4.3 *If there are n first integrals of X_H which are meromorphic and in involution over a neighborhood of the curve $\underline{\Gamma}'$ in M', in particular meromorphic at infinity, and independent in a neighborhood of Γ (not necessarily on Γ itself), then the Galois group of the meromorphic VE over $\bar{\Gamma}'$ is n-involutive Ziglin. Furthermore, the identity component of the Galois group of the VE over $\bar{\Gamma}'$ is abelian.*

We stress that, when we are in the third case, we can consider *three different* Galois groups corresponding to the variational equations over *three*, in general different, Riemann surfaces: Γ, $\bar{\Gamma}$ and $\bar{\Gamma}'$. Each group contains the preceding. But our abelian criterion is less and less precise: the set of *allowed* first integrals is *smaller* at each step. Unfortunately it is in general difficult to compute the differential Galois group. For instance, if the Riemann surface is open it is a transcendental problem. The most favorable situation is when, in the second (respectively the third case), the Riemann surface $\bar{\Gamma}$ (respectively $\bar{\Gamma}'$) is compact and therefore algebraic. Then the connection is defined over a finite extension of the rational functions field $\mathbf{C}(z)$, and there exists an algebraic algorithm to decide if the identity component of the differential Galois group is *solvable*; more precisely, there exists a procedure to find a basis for the space of Liouvillian solutions ([95, 68]). So in that situation we get the existence of a *purely algebraic criterion* (unfortunately not yet effective. . .) for *rational non-integrability*. Notice that, if the manifold M' is algebraic, then $\bar{\Gamma}'$ is algebraic.

It is important to remark that, if the meromorphic VE over $\bar{\Gamma}'$ is *regular singular* (i.e., of Fuchs type), then our three differential Galois groups coincide. Then, if we are in the algebraic situation described above, we get an obstruction not only to the existence of n *rational* first integrals in involution, but more generally to the existence of n first integrals *meromorphic* on the initial manifold M and in involution. In other words, an arbitrary growth at infinity is allowed.

In fact in many practical situations, we have the following: the Riemann surface Γ is an affine curve (i.e., $\Gamma = \Gamma'' - S$, where Γ'' is a compact Riemann surface and S a finite subset), the VE (respectively the NVE) is a holomorphic connection ∇ on a trivial holomorphic bundle over the Riemann surface Γ and it can be extended as a meromorphic connection ∇'' on a trivial bundle over Γ''. Moreover, if this last connection ∇'' is *regular singular*, then the differential Galois groups of ∇ and ∇'' coincide. Therefore we can (theoretically!) compute the differential Galois group algebraically and we can apply Theorem 4.1. Of course we will have in general $\Gamma'' = \bar{\Gamma}$ or $\bar{\Gamma}'$, but, for the applications it is not necessary to verify this fact! These remarks will be very useful for some important non-academic applications. We can conclude that Theorems 4.2 and 4.3 are really interesting when we get *irregular singularities* at the singular points (equilibrium points or points at infinity), in particular in the local situations that we will describe now.

In the following we give two corollaries which are *local* versions of our results.

Locally on Γ or at a regular-singular point of $\bar{\Gamma}$, neither Ziglin's theorem nor our main result are conclusive. But, using our main result, we can get some proofs of local non-integrability at an equilibrium point or at a point at infinity in some cases (see below: Section 4.3, Example 1).

Let X_H be an analytic Hamiltonian system with n degrees of freedom on a symplectic complex manifold M. Let a be an equilibrium point of X_H. Let Γ be a germ of an analytic curve (perhaps singular) at a which is the union of $\{a\}$ and a connected non-stationary germ of the phase curve. Let $\bar{\Gamma}$ be a germ of a smooth holomorphic curve which is a non-singular model for Γ.

Corollary 4.3 *If there are n germs at a of meromorphic first integrals of X_H that are in involution in a neighborhood of a, in particular meromorphic at infinity, and independent in a neighborhood of a (not necessarily on Γ itself), then the local Galois group of the meromorphic germ at a of VE over $\bar{\Gamma}$ is n-involutive Ziglin. Furthermore, the identity component of the local Galois group of the germ at a of VE over $\bar{\Gamma}$ is abelian.*

We add now to the symplectic manifold (M, ω) a hypersurface at infinity M_∞ ($M' = M \cup M_\infty$), and to the curve $i(\Gamma)$ a point at infinity $\infty \in M_\infty$. We get a germ at ∞ of the singular curve Γ' (we suppose that ω admits a meromorphic extension at ∞).

Corollary 4.4 *If there are n germs at ∞ of first integrals of X_H that are meromorphic and in involution in a neighborhood of ∞ in M' and independent in a neighborhood of ∞ (not necessarily on Γ itself), then the local Galois group of the meromorphic germ at ∞ of VE over $\bar{\Gamma}'$ is n-involutive Ziglin. Furthermore, the identity component of the local Galois group of the germ of VE over $\bar{\Gamma}'$ is abelian.*

We now prove Theorem 4.1. Later we will indicate how to modify this proof in order to get Theorems 4.2 and 4.3.

Let V, ∇ be the holomorphic symplectic vector bundle and the connection corresponding to the variational equation of our Hamiltonian system along the solution Γ. On the symmetric bundle S^*V^* of polynomials we can define the structure of a Poisson "algebra" (over the sheaf $\mathcal{O}_\Gamma$ of **C**-algebras of holomorphic functions on Γ) in the following way.

Let d be the differential over the fibre, i.e.

$$d : S^kV^* \longrightarrow S^{k-1}V^* \otimes V^*, \qquad d\alpha = \sum \frac{\partial \alpha}{\partial \eta_i} \otimes \eta_i,$$

where $\eta_1, \ldots, \eta_{2n}$ are fibre coordinates in the bundle V^* (this is a special case of the differential of Spencer).

Then we obtain the mappings,

$$d \otimes d : S^k V^* \otimes S^r V^* \longrightarrow (S^{k-1} V^* \otimes V^*) \otimes (S^{r-1} V^* \otimes V^*)$$

$$Id \otimes \natural : (S^{k-1} V^* \otimes V^* \otimes S^{r-1} V^*) \otimes V^* \longrightarrow S^{k-1} V^* \otimes V^* \otimes S^{r-1} V^* \otimes V^*$$

$$c : S^{k-1} V^* \otimes V^* \otimes S^{r-1} V^* \otimes V \longrightarrow S^{k-1} V^* \otimes S^{r-1} V^*$$

$$\text{sym} : S^{k-1} V^* \otimes S^{r-1} V^* \longrightarrow S^{k+r-2} V^*,$$

where $\natural := \flat^{-1}$, c and sym are the musical isomorphism, the contraction between V and V^*, and the symmetric product, respectively.

The Poisson bracket

$$\{,\} : S^k V^* \otimes S^r V^* \longrightarrow S^{k+r-2} V^*$$

is the composition of the four above maps. It is $\varnothing(\Gamma)$-linear. In a direct way we can prove that it is the usual Poisson bracket in coordinates, if we only derive with respect to the fibre, i.e.

$$\{\alpha, \beta\} = \sum \frac{\partial \alpha}{\partial \eta_i} \frac{\partial \beta}{\partial \xi_i} - \frac{\partial \alpha}{\partial \xi_i} \frac{\partial \beta}{\partial \eta_i},$$

in a canonical frame with canonical coordinates ξ, η. We can extend $\{,\}$ to all the symmetric algebra $S^* V^*$ by bilinearity, and obtain a Poisson algebra $(S^* V^*, \{,\})$ (more precisely a $\mathcal{O}(\Gamma)$-Poisson "algebra").

From now on we fix a point $p_0 \in \Gamma$. Let $E_0 = \text{Sol}_{p_0} \nabla$ be the space of germs of solutions at p_0 (i.e., horizontal vectors of the connection ∇). We can associate to a germ of a solution its initial condition at p_0. We get an isomorphism between E_0 and $E = V_{p_0} = T_{p_0} M$. The $\mathbf{C}$-algebra $\mathbf{C}[E]$ is a complex Poisson subalgebra of the complex Poisson algebra underlying the $\mathcal{O}(\Gamma)$-Poisson algebra $(S^* V^*, \{,\})$, and the natural isomorphism $E_0 \to E$ induces an isomorphism between this Poisson algebra and the natural Poisson algebra $(\mathbf{C}[E], \{,\})$ defined above using the symplectic structure on $E = V_{p_0} = T_{p_0} M$. In the following we identify these two algebras.

The Galois group G of the variational equation acts on $\mathbf{C}[E]$ and the algebra of invariants $\mathbf{C}[E]^G$ is also a Poisson subalgebra. Indeed since G is a symplectic group, G commutes with the symplectic form and, for $\sigma \in G$, $\alpha, \beta \in \mathbf{C}[E]^G$, $\sigma\{\alpha, \beta\} = \{\sigma\alpha, \sigma\beta\} = \{\alpha, \beta\}$.

We can now replace the holomorphic bundle $S^* V^*$, whose sections are functions that are meromorphic on the basis and *polynomial* on the fibre, by the holomorphic locally trivial bundle LV^*, whose sections are functions that are meromorphic on the basis and *rational* on the fibres. We easily extend the preceding constructions to this bundle. The $\mathbf{C}$-algebra $\mathbf{C}(E)$ is a complex Poisson subalgebra of the complex Poisson algebra underlying the $\mathcal{O}(\Gamma)$-Poisson algebra $(LV^*, \{,\})$, and the isomorphism $E_0 \to E$ induces an isomorphism between this Poisson algebra and the Poisson algebra $(\mathbf{C}(E), \{,\})$. As above, we identify these two algebras. The Galois group G of the variational equation acts on $\mathbf{C}(E)$, commutes with the Poisson product, and its algebra of invariants, $\mathbf{C}(E)^G$, is also a Poisson subalgebra.

In the following, by definition, a first integral of the variational equation (or of the corresponding connection ∇) is a meromorphic function defined on the total space of the bundle V, which is meromorphic over the basis, rational over the fibers (i.e., a meromorphic section of the bundle LV^*) and constant on the solutions (i.e., horizontal sections). As the symplectic fibre bundle V is meromorphically trivial as a symplectic bundle, such a first integral can be interpreted as an element of $\mathcal{M}(\Gamma)(\eta_1, \dots, \eta_{2n})$ (in coordinates such that the canonical base, $(1, 0, \dots, 0), \dots, (0, \dots, 0, 1)$, corresponds to a global meromorphic symplectic frame).

Let f be a *holomorphic first integral* defined on a neighborhood of the analytical curve $i(\Gamma)$. Then for any point $p \in \Gamma$ we define the junior part $[f]_p$ of f at p as the first non-vanishing homogeneous Taylor polynomial of f at p with respect to some coordinate system in the phase space. This process has an invariant meaning and the junior part $[f]_p$ must be considered as a homogeneous polynomial on the tangent space $T_p M = V_p$ at p (see[8] for the details). Furthermore, the degree $k \in \mathbf{N}$ of this polynomial is the same for any point $p \in \Gamma$ ([8], Proposition 1.25). In this way, when p varies in Γ, we obtain a holomorphic first integral (polynomial on the fibres) of the variational equation defined on the bundle $T_\Gamma M = V$. It is a holomorphic section of $S^* V^*$.

Let f be now a *meromorphic first integral* defined on a neighborhood of the analytic curve $i(\Gamma)$. Then for any point $p \in \Gamma$ we can naturally extend the map $f \mapsto [f]_p$ to the fraction fields and define the junior part $[f]_p$ of the *meromorphic* first integral f at p. This junior part $[f]_p$ must be considered as a homogeneous rational function on the tangent space $T_p M = V_p$ at p. Furthermore, the degree $k \in \mathbf{Z}$ of this homogeneous rational function is the same for any point $p \in \Gamma$ ([8], Proposition 1.25). In this way, when p varies in Γ, we obtain a meromorphic first integral (rational on the fibres and holomorphic on the basis) of the variational equation defined on the bundle $T_\Gamma M = V$. It is a holomorphic section of LV^*.

Let f, g be two meromorphic first integrals in involution in a neighborhood of the analytic curve $i(\Gamma)$. If we denote respectively by f^0, g^0 the junior parts of them at p_0, then these rational functions are also in involution. Indeed $0 = \{f, g\} = \{f_k + h.o.t., g_r + h.o.t.\} = \{f_k, g_r\} + h.o.t.$, where the first term has the degree $k + r - 2$. The involutivity of f^0 and g^0 follows from this and from the definition of the junior part [8].

Now we recall a fundamental lemma from Ziglin. Let f be a holomorphic function defined over a neighborhood of the origin in a finite dimensional complex vector space E. We define the junior part f^0 of f at the origin [8]. It is a homogeneous element of the rational function field $\mathbf{C}(E)$.

Lemma 4.3 (Ziglin Lemma,[114]) *Let $f_1, \ldots, f_r$ be a set of meromorphic functions over a neighborhood of the origin in the complex vector space E. We suppose that they are (functionally) independent over a punctured neighborhood of the origin (they are not necessarily independent at the origin itself). Then there exist polynomials $P_i \in \mathbf{C}[X_1, \ldots, X_i]$ such that, if $g_i = P_i(f_1, \ldots, f_i)$, then the r rational functions $g_1^0, \ldots, g_r^0 \in \mathbf{C}(E)$ are algebraically independent.*

The following result was proved in [114, 8].

Lemma 4.4 *Let V, ∇ be the holomorphic symplectic vector bundle and the connection corresponding to the variational equation over Γ. Let f^0 be a first integral of the variational equation, holomorphic over the basis and rational over the fibres. Let $p \in \Gamma$. Then the rational function f_p^0 is invariant under the action of the monodromy group $\pi_1(M; p)$.*

The point p defines a representation of the differential Galois group G of the variational equation as a closed (in the Zariski sense) subgroup of $GL(V_p)$. We write $G \subset GL(V_p)$. Then the image $\rho(\pi_1(M; p))$ of the monodromy representation at p is a Zariski dense subgroup of G. We get the following result.

Lemma 4.5 *Let V, ∇ be the holomorphic symplectic vector bundle and the connection corresponding to the variational equation over Γ. Let f^0 be a first integral of the variational equation, holomorphic over the basis and rational over the fibres. Let $p \in \Gamma$. Then the rational function f_p^0 is invariant under the action of the differential Galois group of ∇.*

We need generalizations of this lemma when we have singular points (equilibrium points or points at infinity) and when we consider variational equations over $\bar{\Gamma}$ or $\bar{\Gamma}'$. But in such cases it is in general not true that the image of the monodromy representation is dense in the Galois group and our preceding proof no longer works. Therefore we give below a new proof of Lemma 4.5 which remains valid, mutatis mutandis, in *all cases*. It is very elementary but central in the proof of our main results. In fact we prove a slightly more general result.

Lemma 4.6 *Let V, ∇ be the holomorphic symplectic vector bundle and the connection corresponding to the variational equation over Γ. Let f^0 be a first integral of the variational equation, meromorphic over the basis and rational over the fibres. Let $p \in \Gamma$. We suppose that f^0 is holomorphic over the basis in a neighborhood of p. Then the rational function f_p^0 is invariant under the action of the differential Galois group of ∇.*

We give two different proofs of this lemma.

First Proof. As above we choose a global meromorphic symplectic frame

$$(\phi_1, \dots, \phi_{2n})$$

for the symplectic (meromorphically trivial) holomorphic bundle V over Γ. Using this frame, we identify the field of meromorphic sections of LV^* with

$$\mathcal{M}(\Gamma)(\eta_1, \dots, \eta_{2n}).$$

We fix a point $p \in \Gamma$. The frame allows us to identify the fibre V_p with the space $\mathbf{C}^{2n}$ (with its canonical symplectic structure). We can suppose that $(\phi_1, \dots, \phi_{2n})$ are holomorphic and independent at p. Then, in a neighborhood of p, we can choose a uniformizing variable x over the basis and write the connection ∇ as a differential system $\Delta \eta = \frac{d}{dx}\eta - A(x)\eta$, where A is a holomorphic matrix.

Let $(\zeta_1(x), \dots, \zeta_{2n}(x))$ be the set of solutions satisfying the initial conditions $\zeta_1(0) = (1, 0, \dots, 0), \dots, \zeta_{2n}(0) = (0, \dots, 0, 1)$ at the point p ($x = 0$). The system $(\zeta_1, \dots, \zeta_{2n})$ defines uniquely a Picard-Vessiot extension $\mathcal{M}(\Gamma)\langle \zeta_1, \dots, \zeta_{2n} \rangle$ of the differential field $\mathcal{M}(\Gamma)$. Using this system we get a representation of the differential Galois group G of ∇ as a subgroup of $Sp(2n; \mathbf{C}) \subset GL(V_p)$.

Let $f^0 = f^0(x; \eta_1, \dots, \eta_{2n}) \in \mathcal{M}(\Gamma)(\eta_1, \dots, \eta_{2n})$ be a first integral of the variational equation, meromorphic over the basis and rational over the fibres. We suppose that f^0 is holomorphic over the basis in a neighborhood of p. To a fixed $\lambda = (\lambda_1, \dots, \lambda_{2n}) \in \mathbf{C}^{2n}$ we associate the solution $\eta_\lambda = \lambda_1 \zeta_1 + \dots + \lambda_{2n} \zeta_{2n}$. In a neighborhood of p, we have $f^0(x; \eta_\lambda(x)) = f^0(0; \lambda) \in \mathbf{C}$. We can interpret $f^0(x; \eta_\lambda)$ as an element of the Picard-Vessiot extension $\mathcal{M}(\Gamma)\langle \zeta_1, \dots, \zeta_{2n} \rangle$. In this Picard-Vessiot extension this element is a constant (i.e., it belongs to $\mathbf{C}$), therefore it is invariant under the action of G. When λ varies in $\mathbf{C}^{2n}$ the function $\lambda \mapsto f^0(0; \lambda)$ is rational: it is the expression of the function f_p^0 in coordinates.

Let $\sigma \in G$. Using the definition of a differential Galois group, we get

$$\sigma(f^0(x; \eta_\lambda)) = f^0(x; \sigma(\eta_\lambda)) = f^0(0; \lambda) = f^0(0; \mu),$$

where $\mu = \sigma(\eta_\lambda)(0)$.

Then $\sigma\zeta = B\zeta$, with $B = (b_{ij}) \in Sp(2n; \mathbf{C})$. We have $\sigma(\sum_i \lambda_i \zeta_i) = \sum_{i,j} b_{ij} \lambda_i \zeta_j$. Therefore $\mu = B^t \lambda$ and $f^0(0; \lambda) = f^0(0; B^t \lambda)$. This proves the invariance of f_p^0 under the action of G. $\square$

Second Proof. We sketch a second proof based upon Tannakian arguments. Let $f^0 \in LV^*$ be a first integral of the variational equation. We first suppose that f^0 is not a polynomial. Then we can write $f^0 = \frac{h}{g}$, with $h \in S^k V^*$ and $g \in S^r V^*$ being relatively prime symmetric tensors. If v is a solution of the variational equation (that is $\nabla v = 0$) then the equation $X_h(f^0(v)) = 0$ is equivalent to the equation $(S^k \nabla^* h(v)) g(v) - h(v)(S^r \nabla^* g(v)) = 0$. Consequently we get two equations

$$S^k \nabla^* h = ah,$$

$$S^r \nabla^* g = ag,$$

where a is an element of the coefficient field K of the variational equation.

We set $W = S^k V^* \oplus S^r V^*$ and $\nabla_W = S^k \nabla^* \oplus S^r \nabla^*$. The one-dimensional K-vector subspace $W' = K(h + g)$ of the K-vector space W is clearly stable under the action of the connection ∇_W. We denote by $\nabla_{W'}$ the restriction of ∇_W to W'. Hence we get a rank one subconnection $(W', \nabla_{W'})$ of the connection $(W, \nabla_W) = (S^k V^* \oplus S^r V^*, S^k \nabla^* \oplus S^r \nabla^*)$. This last connection is an object of the tensor category of the (generalized) constructions over ∇. Then we can choose a point $p \in \Gamma$ and introduce the corresponding fibre functor. The space of germs at p of horizontal sections of the subconnection $(W', \nabla_{W'})$ is a complex line in the complex space of horizontal sections of the connection (W, ∇_W). From the Tannakian definition of the Galois group G, this complex line is invariant by G. This complex line is generated over $\mathbf{C}$ by an element $\varphi(h_p + g_p)$ where $\varphi' + a\varphi = 0$. The invariance of the rational function $f_p^0 = \frac{h_p}{g_p}$ by the Galois group G follows immediately. $\square$

The above proof is not far from some arguments used in J.A. Weil's PhD Thesis ([106]) for the study of Darboux's invariants.

We remark that if f^0 is a polynomial, we set $f = f^0$, $g = 1$. The equation $X_h(f^0(v)) = 0$ is equivalent to the equation $(S^k \nabla^* f(v)) = 0$. Then we can replace, in the preceding proof, the connection $(K(h+g), S^k \nabla^* \oplus S^r \nabla^*)$ by the connection $(K(1 + f), \delta_K \oplus S^k \nabla)$. This connection is a rank one subconnection of the connection $(K \oplus S^k V^*, \delta_K \oplus S^k \nabla)$ and we can conclude as above (here the complex line $\mathbf{C}(f_p + 1)$ is invariant by G and G acts trivially on it).

Let now $f_1, \ldots, f_n$ be a family of meromorphic first integrals of the Hamiltonian X_H in involution and independent over a neighborhood of $i(\Gamma)$ (not necessarily on $i(\Gamma)$ itself).

If we apply Ziglin's Lemma (see Lemma 4.3 above) at a point $p \in \Gamma$ to our functions $f_1, \ldots, f_n$, by all the arguments we gave above we get n homogeneous and independent (algebraically and analytically) functions $\alpha_{1_0}, \ldots, \alpha_{n_0}$ in the algebra $\mathbf{C}(E)^G, \{,\}$). In other words, the abelian Lie algebra $(A\{,\})$ of polynomials in $\alpha_{1_0}, \ldots, \alpha_{n_0}$ (with complex coefficients) is a Poisson subalgebra of $\mathbf{C}(E)^G, \{,\}$) and it is invariant by the differential Galois group Γ of the variational equation. Therefore it is annihilated by the Lie algebra $\mathcal{G} = \text{Lie } G$ of the algebraic group G. We complete the proof using Theorem 3.6. This ends the proof of Theorem 4.1. $\square$

With very simple modifications we can obtain the proofs of Theorems 4.2 and 4.3. The essential difference is the following. By hypothesis the first integrals $f_1, \ldots, f_n$ are meromorphic over the manifold M (respectively M'); in particular at the equilibrium points (respectively, at the equilibrium points and at the points at infinity), therefore their junior parts $f_1^0, \ldots, f_n^0$ are *meromorphic* sections over $\bar{\Gamma}$ (respectively $\bar{\Gamma}'$) of the fibre bundle LV^*. Their restrictions over Γ will of course remain holomorphic, but in general they have poles at the singular points and at the points at infinity.

If among our n meromorphic integrals there are some that are functionally independent over Γ, then using the results of Section 4.1, we get

Corollary 4.5 *Let* $f_1, \ldots, f_n$ *be a family of meromorphic first integrals of the Hamiltonian* X_H *in involution and independent over a neighborhood of* $i(\Gamma)$ *(not necessarily on* $i(\Gamma)$ *itself). If moreover, for a fixed integer* $k \leq n$, *the* k *first integrals* $f_1, \ldots, f_k$ *are (functionally) independent over* Γ, *then the Galois group of the NVE is* $n-k$-*involutive Ziglin. Furthermore the identity component of the Galois group of this NVE is abelian.*

There are similar statements in the situation of Theorem 4.2 (respectively 4.3): i.e., when $f_1, \ldots, f_n$ are meromorphic in a neighborhood of $\underline{\Gamma}$ (respectively $\underline{\Gamma}'$). We leave the details to the reader. The proof is essentially the same for all three cases. In order to perform the reduction, we complete $f_1, \ldots, f_n$ into a global meromorphic symplectic frame over Γ (respectively $\bar{\Gamma}$ or $\bar{\Gamma}'$) and we apply the process described above in Section 4.1. We get the NVE. It can have poles, in particular, at the equilibrium points and at the points at infinity. The Galois group G'' of the NVE is a quotient of the Galois group G of the VE. The identity component G^0 of G is abelian, therefore the identity component G'^0 of G' is also abelian. More precisely G^0 is an extension of G''^0, by an algebraic group, isomorphic to some additive group $G_a^r = (\mathbf{C}^r, +)$. We remark that G'^0 can contain a non-trivial torus isomorphic to some multiplicative group $(\mathbf{C}^{*q}, \cdot)$ (cf., our examples below in Section 4.3). So, we observe the possibility of a crucial difference between the first integrals eligible for the normal reduction process and the others.

We remark that, as for the main theorem, the conclusion of the corollary is the same if we restrict the NVE to a neighborhood of some singular point s, and if the Galois group is the local Galois group. In this way we can use our results in order to obtain non-existence of local first integrals in any neighborhood of an equilibrium point or of a point at infinity.

As a final result of this section we show how Ziglin's theorem is a direct consequence of the above corollary when we assume the *complete* integrability of the system, i.e., $n = 2$ in the above corollary.

Corollary 4.6 (Ziglin Theorem, [114]) *Let f be a meromorphic first integral of the two-degrees of freedom Hamiltonian system X_H. We suppose that f and H are independent over a neighborhood of $i(\Gamma)$ (not necessarily on $i(\Gamma)$ itself). Moreover we assume that the monodromy group of the NVE contains a non-resonant transformation g. Then any other transformation belonging to this monodromy group sends eigendirections of g into eigendirections of g.*

Proof. First, we assume that, as in the above results of this section, the set $(i(\Gamma)$ is not reduced to an equilibrium point. Then dH remains different from zero over $i(\Gamma)$ and the reduction to the NVE is made using the one form dH (or in a dual way the vector field X_H).

Then the NVE is given by a symplectic connection over a two-dimensional vector space. Hence its Galois group is an algebraic group whose identity component is abelian and we can identify this group with a subgroup of $SL(2, \mathbf{C})$. In Proposition 2.2, we obtained the classification of the algebraic subgroups of $SL(2, \mathbf{C})$. Here the only possible cases are cases 4 and 5, because for the others either the identity component of the Galois group is not abelian or all the elements of the Galois group are resonant. It is clear that in both cases 4 and 5 we have $g \in G^0$ (we recall that the group topologically generated by a non-resonant element g is a torus, more precisely here this torus is maximal and we have $G^0 \approx \mathbf{C}^*$), and the remaining transformations belonging to the Galois group either preserve the two eigendirections of g or permute them. $\square$

We remark that, as was pointed out by Churchill [22], from our results (essentially from Lemma 4.6, and from the reduction process: Section 4.1) it is possible to prove that we can replace *monodromy group* by *Galois group* in the statement of Ziglin's original theorem (Theorem 3.4).

4.3   Examples

Let X_H be the Hamiltonian system given by the Hamiltonian

$$H = T + U := 1/2(y_1^2 + y_2^2) + 1/2\varphi(x_1) + 1/2\alpha(x_1)x_2^2 + h.o.t.(x_2),$$

with the x_i coordinates and the y_i canonically conjugated momentum, $i = 1, 2$. We assume that the Hamiltonian is holomorphic at the origin.

The plane $\{x_2 = y_2 = 0\}$ is invariant and the Hamiltonian restricted to it is of the type studied in the example of Subsection 4.1. We write $x := x_1$, $y := y_1$. Then we have the integral analytic curve $y^2 + \varphi(x) = 0$. We assume that $\varphi(x) = x^n + h.o.t.$, $n \geq 2$. We want to study the NVE along this integral curve in a neighborhood of the origin (which is an equilibrium point).

The NVE is

$$\frac{d^2\xi}{dt^2} + \alpha(x(t))\xi = 0.$$

As

$$\frac{d}{dt} = (\pm \hat{t}^{n/2} + h.o.t.)\frac{d}{d\hat{t}}$$

for n even, and

$$\frac{d}{dt} = (\frac{1}{2}\hat{t}^{n-1} + h.o.t.)\frac{d}{d\hat{t}}$$

for n odd. We obtain for the corresponding NVE

$$\frac{d^2\xi}{d\hat{t}^2} + (\frac{n}{2\hat{t}} + h.o.t.)\frac{d\xi}{d\hat{t}} + \frac{\alpha(x(\hat{t}))}{\hat{t}^n}\xi = 0, \; n \text{ even,}$$

$$\frac{d^2\xi}{d\hat{t}^2} + (\frac{n-1}{\hat{t}} + h.o.t.)\frac{d\xi}{d\hat{t}} + \frac{4\alpha(x(\hat{t}))}{\hat{t}^{2n-2}}\xi = 0, \; n \text{ odd.}$$

And if

$$\alpha(x) = a_k x^k + h.o.t., \; a_k \neq 0,$$

then, by the Fuchs theorem about regular singular singularities, we get the following result that we state as a proposition for future reference.

Proposition 4.4 *The origin (or more precisely its corresponding points in the desingularized curve) is a regular singular point of the NVE of the above Hamiltonian system along the integral curve $y_1^2 + \varphi(x_1) = 0$ if and only if $n - k \leq 2$, with n and k the multiplicity (as a zero) of $x = 0$ in φ and α, respectively.*

We remark that the above proposition relates the degeneration of the equilibrium points to the irregularity of the corresponding singular points in the variational equation. Hence, the degeneration is related to the (possible) existence of Stokes multipliers.

Now we apply Theorem 2.5 to our Hamiltonian system

$$H = T + U := 1/2(y_1^2 + y_2^2) + 1/2\varphi(x_1) + 1/2\alpha(x_1)x_2^2 + h.o.t.(x_2),$$

(two degrees of freedom), with the integral irreducible analytic curve defined by $y^2 + \varphi(x) = 0$ (as above, we drop the subindexes). Furthermore we assume that φ and α are polynomials. Then $\overline{\Gamma}$ is a compact Riemann surface (see [53]) and the usual change of variables $x \leftrightarrow t$, $x = x(t)$ ($x(t)$ is the solution of the hyperelliptic differential equation $\dot{x}^2 + \varphi(x) = 0$) gives us a pull-back of the NVE on the Riemann sphere (the classics call it the algebraic form of the equation, [109, 88])

$$\frac{d^2\xi}{dx^2} + \frac{\varphi'(x)}{2\varphi(x)}\frac{d\xi}{dx} - \frac{\alpha(x)}{2\varphi(x)}\xi = 0.$$

We call this equation (ANVE): the algebraic NVE (as in other chapters).

We observe that the singular points are the branching points of the covering (i.e., the roots of φ and the point at infinity). Concerning the equilibrium points of the original Hamiltonian, we see that $x = 0$ is a singular point if $n - k > 0$ and it is irregular if $n - k > 2$ in complete accordance with the last proposition.

Furthermore by Theorem 2.5, the identity components of the Galois groups of the NVE and of the ANVE are the same. We now look at the standard transformation in order to put the ANVE in the normal invariant form

$$\frac{d^2\xi}{dx^2} + I(x)\xi = 0, \qquad \text{with} \qquad I := q - \frac{1}{4}p^2 - \frac{1}{2}\frac{dp}{dx},$$

and

$$\frac{d^2\xi}{dx^2} + p\frac{d\xi}{dx} + q\xi = 0,$$

the original equation, where we conserve the symbol ξ for the new variable. Note that we have introduced an algebraic function $(\exp(-1/2\int p = \varphi^{-1/4})$ only. Hence, the identity components of the Galois groups of the ANVE and of its normal invariant form are the same, and we can identify the two equations in order to obtain non-integrability results. As a conclusion, we can work directly with the normal form of the ANVE: if the identity component of its Galois group is not abelian, the Hamiltonian system is not integrable (we observe that the Galois group of the normal invariant form and the Galois group of the initial NVE as well, are contained in $SL(2, \mathbf{C})$: see Section 2.2).

Example 1. We apply the theory to the very simple Hamiltonian of two degrees of freedom with two parameters,

$$H = T + U = \frac{1}{2}(y_1^2 + y_2^2) + \frac{1}{3}x_1^3 + \frac{1}{2}(a + bx_1)x_2^2, \ a \in C^*, b \in C.$$

This Hamiltonian system has the integral curve,

$$\Gamma : \quad \dot{x}_1^2 = -\frac{2}{3}x_1^3, \quad x_1 = -6t^{-2}, \quad y_1 = 12t^{-3}, \quad x_2 = y_2 = 0.$$

The corresponding normal variational equation is

$$\ddot{\xi} + (a - 6bt^{-2})\xi = 0.$$

We observe that there are two singular points, the origin and the point at infinity. The first one is irregular (by the above proposition) and the second one regular singular. In fact, as we shall see, the NVE is a confluent hypergeometric equation.

By making the change of variables $t = \frac{iz}{2\sqrt{a}}$, we get

$$\frac{d^2\xi}{dz^2} - (\frac{1}{4} + 6b\frac{1}{z^2})\eta = 0.$$

This equation is a family of Whittaker equations, with one parameter only. In fact, as we know from Subsection 2.8.4, it can be transformed into a family of Bessel equations. Then the identity component of the Galois group of the NVE is abelian if, and only if, $\mu + 1/2$ is integer. Hence, by Corollary 4.5, for $b \neq \frac{1}{6}(k^2 + k)$, $k \in \mathbf{Z}$, this Hamiltonian system is not integrable (i.e., it does not have a global meromorphic first integral beyond the Hamiltonian).

We observe that for $a = 0$ the above Hamiltonian is the homogeneous Hénon-Heiles Hamiltonian. It will be studied from the differential Galois point of view in Chapters 5 and 6.

Example 2. For three degrees of freedom one has the following natural generalization of Example 1,

$$H = T + U = \frac{1}{2}(y_1^2 + y_2^2 + y_3^2) + \frac{1}{3}x_1^3 + \frac{1}{2}(A + Bx_1)x_2^2 + \frac{1}{2}(C + Dx_1)x_3^2,$$

where $A, C \in \mathbf{C}^*$, $B, D \in \mathbf{C}$.

This Hamiltonian system has the integral curve,

$$\Gamma: \quad \dot{x}_1^2 = -\frac{2}{3}x_1^3, \quad x_1 = -6t^{-2}, \quad y_1 = 12t^{-3}, \quad x_2 = x_3 = y_2 = y_3 = 0.$$

The corresponding NVE is composed of two independent Whittaker equations of the same type (at the level of connections it is a direct sum of two connections, each of them being the connection of a Whittaker equation). Hence, if one of the parameters B or D is different of $\frac{1}{6}(k^2 + k)$, $k \in \mathbf{Z}$, we get non-integrability. In the same way we can generalize this to more degrees of freedom and to some other examples when the NVE is split in 2×2 equations of the same kind.

Example 3. We consider now the family of Hamiltonian systems with two degrees of freedom defined as above with $\varphi(x) = x^n$, $\alpha(x) = ax^{n-4} + bx^{n-3} + cx^{n-2}$, where n is an integer, $n > 3$ and a, b, c are complex parameters, with $a \neq 0$.

In the same situation as above, the normal invariant form of the ANVE is

$$\frac{d^2\xi}{dx^2} - ((\frac{n(n-1)}{16} + c)x^{-2} + bx^{-3} + ax^{-4})\xi = 0.$$

With the change of variables $x = \frac{\hat{x}}{2\sqrt{a}}$, we get

$$\frac{d^2\xi}{dx^2} - (\frac{1}{4a}(\frac{n(n-1)}{16} + c)x^{-2} + \frac{b}{4a}x^{-3} + \frac{1}{4}x^{-4})\xi = 0,$$

where in order to simplify the notation we write again x instead of $\hat{x}$). Now, if
in the Whittaker equation

$$\frac{d^2\xi}{dz^2} - (\frac{1}{4} - \frac{\kappa}{z} + \frac{4\mu^2 - 1}{4z^2})\eta = 0,$$

we do the change of variables, $z = 1/x$, we obtain

$$\frac{d^2\xi}{dx^2} - (\frac{4\mu^2 - 1}{4}x^{-2} - \kappa x^{-3} + \frac{1}{4}x^{-4})\xi = 0.$$

So, the ANVE is a general Whittaker equation, with

$$4\mu = \sqrt{c + \frac{n(n-4)}{16} + \frac{1}{4}}, \quad \kappa = -\frac{b}{2\sqrt{a}}.$$

We recall that if

$$p := \kappa + \mu - \frac{1}{2} \qquad q := \kappa - \mu - \frac{1}{2},$$

then, the identity component of the Galois group of the Whittaker equation is
abelian if, and only if, (p, q) belong to $(\mathbf{N} \times -\mathbf{N}^*) \cup (-\mathbf{N}^* \times \mathbf{N})$, i.e., p, q are
integer, one of them positive and the other negative. Hence, this last condition
is a necessary condition for the integrability of the initial Hamiltonian system.

We make two remarks about the above Examples 1–3. The first one is that,
because of abelian character of the monodromy group of the NVE, we can not
obtain any non-integrability result by an analysis of the monodromy group. The
second one is that, as the NVE are confluent hypergeometric equations, the *local*
Galois group at the irregular point and the *global* Galois group coincide. Then,
by Corollary 4.3 we have proved indeed the non-integrability of these systems
in any neighborhood of the origin in the complex phase space (the equilibrium
point corresponding to the irregular singular point).

For some special Hamiltonians it is also possible to prove local non-
integrability in a Fuchsian context as is shown by Ziglin in the following example
that we include for the sake of completeness.

Example 4 ([114]). We recall briefly the Ziglin analysis. The starting Hamiltonian is

$$H = \frac{1}{2}(y_1^2 + y_2^2 + x_1^2 x_2^2).$$

By an elementary canonical transformation (a rotation, which puts one
of the symmetric invariant planes on the axis $x_2 = y_2 = 0$), Ziglin obtained a
Hamiltonian system with potential

$$V(x_1, x_2) = \frac{1}{8}(x_1^2 - x_2^2)^2,$$

where we keep the same notation for the new coordinates.

This is a potential of the type studied above with $\varphi(x_1) = 1/4x_1^4$ and $\alpha(x_1) = -1/2x_1^2$.

Then Ziglin considered the NVE along the family of integral curves $x_2 = y_2 = 0$ using as a parameter the energy $H > 0$ (he did not consider the integral curve through the origin). These variational equations are reduced to Lamé's type and then he applies his theorem about the monodromy group (in fact, by the proof of Corollary 4.6, he proved the non-abelian structure of the identity component of the Galois Group). So the system under study does not have an additional meromorphic first integral in a neighborhood of the above family of integral curves.

The key point now is that by the (quasi-) projective structure of the Hamiltonian (the potential is a homogeneous polynomial) if the system has a holomorphic first integral, then any of the homogeneous polynomials in the expansion of this integral at the origin must also be a first integral. In this way Ziglin proved the local non-integrability of this system at the origin.

In the next chapter we shall study a family of Hamiltonian systems that generalize this last example (see the Umeno families in Subsection 5.1.3).

Chapter 5

Three Models

We start the part devoted to applications with three important non-academic models: the homogeneous potentials, the Bianchi IX cosmological model and the Sitnikov system in celestial mechanics. We note that, from the differential Galois theory of Chapter 2 (we shall need only the theorem of Kimura and the algorithm of Kovacic) and from our results of Chapter 4, the methods proposed here are completely systematic and elementary. In our opinion, this reflects the fact that the natural setting to obtain non-integrability results, using an analysis of the variational equations (along a particular integral curve), is the differential Galois theory.

The results of this chapter were obtained in a joint work of the author with J.P. Ramis [78, 79], except Example 3 of Subsection 5.1.3 is new and will be also studied in Chapter 6 from a different point of view.

5.1 Homogeneous potentials

5.1.1 The model

The purpose of this section is to give a simple non-integrability criterion for complex Hamiltonian systems with homogeneous potentials, i.e., of the type

$$H(x^c\, y) = T + V = \frac{1}{2}(y_1^2 + \cdots + y_n^2) + V(x_1, \ldots, x_n),$$

where V is a homogeneous function of integer degree k. We consider this as a first non-completely academic application in order to test the results of Chapter 4. In this way, we show that we can improve some of the Yoshida results even for two degrees of freedom (see [111]), avoiding arithmetical problems related with the non-resonance assumptions in Ziglin's Theorem or its generalizations,

J. J. Morales Ruiz, *Differential Galois Theory and Non-Integrability of Hamiltonian*, Systems, Modern Birkhäuser Classics,
DOI: 10.1007/978-3-0348-0723-4_5, © Springer Basel 1999

because these theorems are exclusively based on the analysis of the monodromy group of the variational equations (see [101]).

We take three concrete examples: the collinear homogeneous problem of three particles studied by Yoshida in [111], the n-degrees of freedom system with potential

$$V = \frac{1}{s} \sum x_{i_1}^s x_{i_2}^s \cdots x_{i_r}^s,$$

taken by Umeno [101] and the so-called Hénon-Heiles homogeneous potential studied by Yoshida [111].

5.1.2   Non-integrability theorem

Let an n degree of freedom Hamiltonian system with Hamiltonian

$$H(x, y) = T + V = \frac{1}{2}(y_1^2 + \cdots + y_n^2) + V(x_1, \ldots, x_n), \qquad (5.1)$$

V being a homogeneous function of integer degree k and $2 \leq n$. For the case $n = 2$, Yoshida obtained a remarkable non-integrability theorem based on Ziglin's Theorem [111].

As Yoshida notes, it is not possible to generalize, in a direct way, his theorem to $n > 2$. Indeed, it is difficult to check the non-resonant condition of Ziglin's Theorem. He asks for a generalization of Ziglin's theorem in order to handle these systems ([111], p. 141). With Theorem 5.1 it is possible to solve this problem and moreover, we can improve Yoshida's results, even for the case $n = 2$.

We follow the Yoshida arguments in order to obtain a set of hypergeometric equations as the NVE along a particular solution of the Hamiltonian system $H(x, y)$ above. From the homogeneity of V, it is possible to get an invariant plane

$$x = z(t)c_i, \; y = \dot{z}(t)c_i, \; i = 1, 2, \ldots, n,$$

where $z = z(t)$ is a solution of the (scalar) hyperelliptic differential equation

$$\dot{z}^2 = \frac{2}{k}(1 - z^k),$$

(we assume the non-trivial case $k \neq 0$) and $c = (c_1, c_2, \ldots, c_n)$ is a solution of the equation

$$c = V'(c).$$

This is the particular solution Γ along which we compute the VE and the NVE.

The VE along Γ is given, in the temporal parametrization, by

$$\ddot{\eta} = -z(t)^{k-2} V''(c)\eta.$$

By the symmetry of the Hessian matrix $V''(c)$, it is possible to express the VE as a direct sum of second order equations

$$\ddot{\eta}_i = -z(t)^{k-2}\lambda_i\eta_i, \; i = 1, 2, \ldots, n,$$

where we preserve η for the new variable, λ_i being the eigenvalues of the matrix $V''(c)$. We call these eigenvalues Yoshida coefficients. One of the above second order equations is the tangential variational equation, say, the equation corresponding to $\lambda_n = k - 1$. This equation is trivially integrable and we get as NVE an equation in the variables $\xi := (\eta_1, \ldots, \eta_{n-1}) := (\xi_1, \ldots, \xi_{n-1})$, i.e.,

$$\ddot{\xi} = -z(t)^{k-2}\operatorname{diag}(\lambda_1, \ldots, \lambda_{n-1})\xi.$$

Now, following Yoshida, we consider the finite branched covering map

$$\overline{\Gamma} \to \mathbf{P}^1,$$

given by $t \mapsto x$, with $x =: z(t)^k$ (here $\overline{\Gamma}$ is the compact hyperelliptic Riemann surface of the hyperelliptic curve $w^2 = \frac{2}{k}(1 - z^k)$, see Subsection 4.1.1). By the symmetries of this problem, we get as NVE a system of independent hypergeometric differential equations in the new independent variable x

$$x(1-x)\frac{d^2\xi}{dx^2} + (\frac{k-1}{k} - \frac{3k-2}{2k}x)\frac{d\xi}{dx} + \frac{\lambda_i}{2k}\xi = 0, \; i = 1, 2, \ldots, n-1.$$

This system of equations is the algebraic normal variational equation ($ANVE$). If we write $ANVE_i$ for the scalar second order equation corresponding to the Yoshida coefficient λ_i then

$$ANVE = ANVE_1 \oplus ANVE_2 \oplus \cdots \oplus ANVE_{n-1},$$

(in fact it is a direct sum in the more intrinsic sense of linear connections of Chapter 2). Then it is clear that the $ANVE$ is integrable if, and only if, each one of the $ANVE_i$'s is integrable. In other words, the identity component of the Galois Group of the $ANVE$ is solvable if, and only if, each one of the identity components of the Galois Group of the $ANVE_i$'s, $i = 1, 2, \ldots, n-1$, is solvable.

Each one of the above $ANVE_i$'s is a hypergeometric equation with three regular singular points at $x = 0$, $x = 1$ and $x = \infty$. We remark that, by Theorem 2.5, the identity component of the Galois Group of the NVE is the same as the identity component of the Galois Group of the $ANVE$.

For the equation $ANVE_i$, the exponent differences at $x = 0$, $x = \infty$ and $x = 1$ are, respectively, $\hat{\lambda} = 1/k$, $\hat{\mu} = \sqrt{(k-2)^2 + 8k\lambda_i}/(2k)$ and $\hat{\nu} = 1/2$. Now we obtain the following result.

Theorem 5.1 *If the Hamiltonian system with Hamiltonian (5.7) is completely integrable (with holomorphic or meromorphic) first integrals, then each pair (k, λ_i) belongs to one of the following lists (we do not consider the trivial case $k = 0$)*

(1) $(k, p + p(p-1)k/2)$,

(2) $(2,$ *arbitrary complex number*$)$,

(3) $(-2,$ *arbitrary complex number*$)$,

(4) $(-5, \frac{49}{40} - \frac{1}{40}(\frac{10}{3} + 10p)^2)$,

(5) $(-5, \frac{49}{40} - \frac{1}{40}(4 + 10p)^2)$,

(6) $(-4, \frac{9}{8} - \frac{1}{8}(\frac{4}{3} + 4p)^2)$,

(7) $(-3, \frac{25}{24} - \frac{1}{24}(2 + 6p)^2)$,

(8) $(-3, \frac{25}{24} - \frac{1}{24}(\frac{3}{2} + 6p)^2)$,

(9) $(-3, \frac{25}{24} - \frac{1}{24}(\frac{6}{5} + 6p)^2)$,

(10) $(-3, \frac{25}{24} - \frac{1}{24}(\frac{12}{5} + 6p)^2)$,

(11) $(3, -\frac{1}{24} + \frac{1}{24}(2 + 6p)^2)$,

(12) $(3, -\frac{1}{24} + \frac{1}{24}(\frac{3}{2} + 6p)^2)$,

(13) $(3, -\frac{1}{24} + \frac{1}{24}(\frac{6}{5} + 6p)^2)$,

(14) $(3, -\frac{1}{24} + \frac{1}{24}(\frac{12}{5} + 6p)^2)$,

(15) $(4, -\frac{1}{8} + \frac{1}{8}(\frac{4}{3} + 4p)^2)$,

(16) $(5, -\frac{9}{40} + \frac{1}{40}(\frac{10}{3} + 10p)^2)$,

(17) $(5, -\frac{9}{40} + \frac{1}{40}(4 + 10p)^2)$,

(18) $(k, \frac{1}{2}(\frac{k-1}{k} + p(p+1)k))$,

where p is an arbitrary integer.

Proof. The proof follows from our Corollary 4.1 and from Kimura's theorem, because if the identity component of the Galois group is abelian, then, in particular, it is solvable. Case (1) corresponds to case (i) in Kimura's theorem and cases (2)–(18) to case (ii) in Kimura's theorem. In particular, in cases (4)–(18) the Galois Group is finite and the identity component of the Galois Group (of the *ANVE* and of the *NVE*) is trivial.   □

We recall Yoshida's theorem. For $n = 2$ (only one parameter λ appears), let us consider the four regions

(i) $S_k = \{\lambda > 1, -j(j+1)|k|/2 - j + 1 > \lambda > -j(j+1)|k|/2 + j + 1, j \in \mathbf{N}\}$, for $k \leq -3$,

(ii) $S_{-1} = \mathbf{C} - \{-j(j-1)/2 + 1, j \in \mathbf{N}\}$,

(iii) $S_1 = \mathbf{C} - \{j(j-1)/2 + 1, j \in \mathbf{N}\}$,

(iv) $S_k = \{\lambda < 0, j(j-1)k/2 + j < \lambda < j(j+1)k/2 - j, j \in \mathbf{N}\}$, for $k \geq 3$.

Then we have [111]

Theorem 5.2 *If λ is in the region S_k then the corresponding Hamiltonian system is not integrable.*

Now, it is easy to show that Yoshida's theorem is a particular case of Theorem 5.1. We sketch the steps. For case (1) of Theorem 5.1, we see that the parameter λ belongs to the complement, in the complex plane, of the Yoshida non-integrability regions S_k, $k \in \mathbf{Z} - \{2, -2\}$, because $S_1 = \mathbf{C} - \{p + p(p-1)/2\}$, $S_{-1} = \mathbf{C} - \{p - p(p+1)/2\}$ (with $p \in \mathbf{Z}$), and for the rest of the k values, $\lambda = p + p(p-1)k/2$ are precisely the extremities of the open intervals that appear in S_k. For the cases (3)–(18) we give an indirect argument. These cases correspond to a finite Galois (and monodromy) Group, and Yoshida's theorem is based, as Ziglin's theorem, on the existence of a non-resonant monodromy matrix, but for these cases all the Galois (and monodromy) transformations of the *NVE* are resonant, then necessarily the values of the Yoshida coefficient λ for the cases (3)–(18) are contained in $\mathbf{C} - \cup S_k$. We recall that, by Theorem 2.5, the identity component of the Galois group is preserved when we obtain the *ANVE* from the *NVE* and that a linear algebraic group is finite if and only if the identity component is trivial.

The knowledgeable reader may ask why we do not use the last argument (used for cases (3)–(18)) for the case (2). The reason is that, in this case, the Galois group might have an identity component solvable but not abelian, and in Yoshida's (and Ziglin's) theorem, a necessary condition for integrability is the abelianess of the identity component of the Galois Group. In fact, as was proved in Chapter 4 for a two degrees of freedom Hamiltonian system, Ziglin's theorem is a consequence of our main results of Chapter 4, see Corollary 4.6.

5.1.3 Examples

Example 1. We consider with Yoshida the collinear three body problem with a homogeneous potential

$$V(q_1, q_2, q_3) = |q_1 - q_2|^k + |q_1 - q_3|^k + |q_2 - q_3|^k.$$

By a reduction to the center of masses it is transformed to a two degrees of freedom potential

$$V(x_1, x_2) = (\sqrt{3}x_1 + x_2)^k + (-\sqrt{3}x_1 + x_2)^k + (2x_2)^k.$$

For k an arbitrary integer, it is possible to obtain a hyperelliptic integral curve with *ANVE* having Yoshida parameter $\lambda = 3(k-1)/(1 + 2^{k-1})$. For k even and positive there exists an additional hyperelliptic integral curve with an *ANVE* having $\lambda = (k-1)/3$ [111].

By applying the last theorem we get

Proposition 5.1 *Except for the four cases* $k = -2, 1, 2, 4$, *the collinear homogeneous potential of three particles of degree* k *is not integrable (we do not consider the trivially integrable case* $k = 0$*).*

We remark that, with respect to Yoshida's results, the new feature we find is the non-integrability of the system for k odd and greater than or equal to 5. Furthermore, the four cases $k = -2, 1, 2$ and 4 are well-known integrable systems (see the references in [111]). In this way we close the integrability problem for this family.

Example 2. We now apply Theorem 5.1 to a family of systems with an arbitrary number of degrees of freedom, studied by Umeno [101].

In order to avoid the already mentioned arithmetical problems (for n greater than two) related to the non-resonance condition, Umeno introduced the non-resonance-degenerate condition. In this way he studied the non-integrability of the systems given by the very symmetric n-degrees of freedom homogeneous potentials

$$V = \frac{1}{s} \sum x_{i_1}^s x_{i_2}^s \cdots x_{i_r}^s,$$

where the summation considers all the possible combinations of r different integers, $i_1, i_2, \ldots, i_r$, with i_p equal to $1, 2, \ldots, n$. Following Umeno, we denote this Hamiltonian system by (n, r, s). We observe that, in any case, $r \leq n$.

For these systems, we have $k = rs$ and it is possible (see [101]) to find a hyperelliptic curve with associated $ANVE$ admitting as Yoshida parameters the values

$$\lambda_1 = \lambda_2 = \cdots = \lambda_{n-1} = s - 1 - \frac{s(r-1)}{n-1} := \lambda.$$

Then, by Theorem 5.1, we get

Proposition 5.2 *The above Hamiltonian systems with parameters* (n, r, s) $(2 \leq n)$ *are not completely integrable, except perhaps for the five cases*

(i) $n(s - 1) = rs - 1$,

(ii) $n(s - 2) = rs - 2$,

(iii) $r = 1$,

(iv) $rs = 2$

(v) $2 \left(s - 1 - \frac{s(r-1)}{n-1} \right) = \frac{sr-1}{sr}$.

Proof. We need to check cases (1), (2) and (11)–(18) of Theorem 5.1. Case (iv) corresponds to case (2). Now, it is easy to see that $\lambda = s - 1 - \frac{s(r-1)}{n-1} \leq rs - 1$ (this fact is also used in [101]). Hence, it is only necessary to consider the values $p = 0, 1$ and -1 in case (1) of Theorem 5.1 (for other values of p we do not have positive integer values of n, r and s). These values give us conditions (i), (ii), (iii) and (v). This last case corresponds to (18) with $p = 0$.

Case (11) is not possible because $rs = 3$ implies $r = 1$, $s = 3$ or $r = 3$, $s = 1$. As $r = 1$ appears in (iii), it is only necessary to consider $r = 3$, $s = 1$. But the equation $-2/(n-1) = -1/24 + (2 + 6p)^2/24$ has no solutions $n \in \mathbf{N}$ and $p \in \mathbf{Z}$. In a similar way, cases (12)–(17) are not possible. □

We observe that the trivial case (iii) corresponds to a separable potential.

In his paper Umeno proved the *non-existence of an additional integral* for the following systems:

(a) $(n, r, 1)$ with $3 \leq r$,

(b) $(n, r, 2)$ except for the five cases $(2, 2, 2)$, $(3, 3, 2)$, $(26, 4, 2)$, $(6, 6, 2)$, $(28, 6, 2)$, the two families $(2r - 1, r, 2)$, $(r + 1, r, 2)$ $(2 \leq r)$ and the trivial family $r = 1$,

(c) (n, r, s) with $2 < s$, $\frac{rs-2}{s-2} < n$ and $2 \leq r$.

We remark that condition (c) is incompatible with condition (v) of Proposition 5.2. We first notice that $n - 1 > 0$. Then condition (v) implies $2(s - 1 - \frac{s(r-1)}{n-1}) = \frac{sr-1}{sr} < 1$. Therefore $2(s - 1)(n - 1) - 2sr + 2s < n - 1$ and $2s(n - r) < 3(n - 1)$. From condition (c) we get $s > 2$ and $rs - 2 < sn - 2n$. Therefore $2s(n - r) > 4(n - 1)$. Finally $3(n - 1) > 4(n - 1)$, $0 > n - 1$, and we get a contradiction.

Using similar (even simpler) arguments, starting from Proposition 5.2 we can prove the *non-complete integrability* of systems (a), (b) and (c) and, furthermore, we get new non-integrable systems. For instance, the systems of type $(n, r, 2)$ $(2 \leq n)$ are not completely integrable except for $(n, 1, 2)$ and $(2r - 1, r, 2)$. Umeno considers in [101] as an open problem for these systems, with $r = 2$, the fact that among the systems in (b) there are some that have a non-integer difference of Kowalevski exponents, but his criterion can not decide their non-integrability. We have proved above that all these systems are never completely integrable.

We note that the non-integrable system with parameters $(2, 2, 2)$ already appeared in this monograph (see Section 4.3): it is the Yang-Mills potential studied by Ziglin.

Example 3. The homogeneous Hénon-Heiles Hamiltonian is given by the Hamiltonian function

$$H = \frac{1}{2}\left(y_1^2 + y_2^2\right) + \frac{e}{3}x_1^3 + x_1 x_2^2. \tag{5.2}$$

The parameter e is assumed to be complex. We notice, first, that by a suitable scaling the potential can also be taken as $\frac{1}{3}x_1^3 + \frac{1}{e}x_1 x_2^2$. So, the case $e = \infty$ is integrable. Furthermore, a rotation in the configuration space (eventually, a complex rotation) converts (5.2) into

$$H = \frac{1}{2}(\eta_1^2 + \eta_2^2) + \frac{2}{3(e-1)}\xi_1^3 + \xi_1\xi_2^2 + \frac{\sqrt{2-e}}{3}\frac{e+1}{e-1}\xi_2^3, \qquad (5.3)$$

if $e \neq 1$. A suitable scaling has also been used. If $e = 1$, it reduces to $\frac{1}{2}(\eta_1^2 + \eta_2^2) + \frac{1}{3}(\xi_1^3 + \xi_2^3)$, which is clearly separable. Skipping the $O(\xi_2^3)$ terms, we remark that (5.3) is as (5.2) with $\hat{e} = \frac{2}{e-1}$.

In [111] Yoshida computed the values of the corresponding Yoshida's parameters as $\lambda = \frac{2}{e}$, $\hat{\lambda} = \hat{e} - 1$, and by application of his theorem he obtained the following result.

Proposition 5.3 ([111]) *The Hamiltonian system with Hamiltonian* (5.2) *is non-integrable for* $e \in (-\infty, 1) \cup \left(\bigcup_{j=1}^{\infty}(1 + j + 3\binom{j}{2}, 1 - j + 3\binom{j+1}{2})\right)$.

We observe that the complement of this set of values of e contains infinitely many intervals, with increasing lengths.

The use of the Painlevé property (see [89]) suggests that the system (5.2) is integrable only for $e = 1, 6, 16$ and some difficulties appear in the case $e = 2$ (see [38]).

Beyond the separable case $e = 1$, the cases $e = 6$ and $e = 16$ are known to be integrable. We display the first integrals independent of the Hamiltonian [89]:

1. For $e = 6$: $F = 4x_1^2 x_2^2 + x_2^4 - 4x_1 y_2^2 + 4x_2 y_1 y_2$,
2. For $e = 16$: $F = y_2^4 + 4x_1 x_2^2 y_2^2 - \frac{4}{3}x_2^3 y_1 y_2 - \frac{4}{3}x_1^2 x_2^4 - \frac{2}{9}x_2^6$.

By application of Theorem 5.1 we get the following non-integrability result.

Proposition 5.4 *The Hamiltonian* (5.2) *is non-integrable for*

$$e \in \mathbf{C}\backslash\{1, 2, 6, 16\}.$$

Proof. From Theorem 5.1, the Yoshida parameter (λ and $\hat{\lambda}$) can take the following six possible values: $f_1(p) := p + \frac{3}{2}p(p-1)$, $f_2(p) := \frac{1}{3} + \frac{3}{2}p(p-1)$, $f_3(p) := -\frac{1}{24} + \frac{1}{24}(2 + 6p)^2$, $f_4(p) := -\frac{1}{24} + \frac{1}{24}(\frac{3}{2} + 6p)^2$, $f_5(p) := -\frac{1}{24} + \frac{1}{24}(\frac{6}{5} + 6p)^2$, $f_6(p) := -\frac{1}{24} + \frac{1}{24}(\frac{12}{5} + 12p)^2$, $p \in \mathbf{Z}$. The integrability of the Hamiltonian (5.2) is only compatible with the identity $\frac{2}{\lambda} = e = \hat{\lambda} + 1$, where λ and $\hat{\lambda}$ take values in the above six families $f_i(p)$, $i = 1, \ldots, 6$.

It is clear that $\hat{\lambda} + 1 \geq 1$. On the other hand, we have $2/f_1(p) \leq 2$, $2/f_2(p) \leq 6$, $2/f_3(p) \leq 16, 2/f_4(p) \leq 192/5, 2/f_5(p) \leq 1200/11$ and $2/f_6(p) \leq 1200/119$. Then we only need to check a finite number of cases and the only ones of these that verify the above identity are for $e = 1, 2, 6, 16$. $\qquad\square$

We shall give a different proof of the above proposition in the next chapter.

This example was suggested to the author by Ljubomir Gavrilov some years ago.

As a final remark, we think that it is possible to apply Theorem 5.1 to other interesting systems.

5.2 The Bianchi IX cosmological model

5.2.1 The model

The Bianchi IX Cosmological model is a dynamical system given by the equations in logarithmic time [60]

$$\frac{d^2 \log x_1}{dt^2} = (x_2 - x_3)^2 - x_1^2, \qquad \frac{d^2 \log x_2}{dt^2} = (x_3 - x_1)^2 - x_2^2, 1$$

$$\frac{d^2 \log x_3}{dt^2} = (x_1 - x_2)^2 - x_3^2, 1 \tag{5.4}$$

with the energy constraint (from physical considerations)

$$E = -\left(\frac{\dot{x}_1}{x_1}\frac{\dot{x}_2}{x_2} + \frac{\dot{x}_2}{x_2}\frac{\dot{x}_3}{x_3} + \frac{\dot{x}_3}{x_3}\frac{\dot{x}_1}{x_1}\right) + x_1^2 + x_2^2 + x_3^2 - 2(x_1 x_2 + x_2 x_3 + x_3 x_1) = 0.$$

So, we get a dynamical system of dimension five on the zero level energy manifold M_0.

In fact this system is a Hamiltonian system with positions x_1, x_2, x_3 and their conjugate moments given by

$$y_1 = -\frac{1}{x_1}\frac{d}{dt}\log(x_2 x_3), \quad y_2 = -\frac{1}{x_2}\frac{d}{dt}\log(x_1 x_3), \quad y_3 = -\frac{1}{x_3}\frac{d}{dt}\log(x_1 x_2).$$

So that the energy becomes the Hamiltonian

$$H = \frac{1}{4}(x_1^2 y_1^2 + x_2^2 y_2^2 + x_3^2 y_3^2 - 2x_1 x_2 y_1 y_2 - 2x_2 x_3 y_2 y_3 - 2x_1 x_3 y_1 y_3)$$

$$+ x_1^2 + x_2^2 + x_3^2 - 2x_1 x_2 - 2x_2 x_3 - 2x_1 x_3 = 0.$$

This system is studied from a real dynamical point of view in [60, 13, 73, 28] (the real configuration space for this system is given by $x_i > 0$, $i = 1, 2, 3$). It is proved in [28] that it is a locally integrable Hamiltonian system (for a precise definition of what that means, see the above reference) on the open real physical phase space. As it is not recurrent in the real open phase space, it is necessary to compactify this physical phase space in order to have recurrence and then it is proved in [13] (p. 54–76) that the behaviour of the system is very complicated in this extended real phase space (this is also studied in a more heuristic way in [60], p. 477–484) with oscillations around the gravitational collapse (the gravitational collapse is given by the points in the phase space with $x_1 x_2 x_3 = 0$). It seems that some insight into this *real* chaotic behaviour of the Bianchi IX model was also recently given by Cornish and Levin in the preprint [20] (this and other additional recent references on the dynamics of Bianchi IX are given in [11]).

From the complex dynamics (with *complex* time) approach more recently the paper [61] is devoted to showing that this dynamical system has not passed the so-called Painlevé test (in fact, as remarked in Chapter 1, this method started with the Kowaleski analysis of the rigid body [57]), i.e., that the only movable singularities of the solutions with respect to the time parametrization are poles (sometimes the existence of movable essential singularities is also permitted). The authors used the variational equations along a family of particular solutions founded by Taub [99], and they showed that these variational equations have an irregular singular point and, hence, the system "contains" essential singularities in their solutions around the Taub family. We notice that from the existence of an irregular singularity it follows immediately that the general solution of this system (with five parameters) is not meromorphic in time.

From the above considerations, it is clear that the Bianchi IX model is a polemic system and apparently contradictory results were obtained. Part of this puzzle is clarified by the following two remarks. We recall, as remarked in Chapter 3, that there are several degrees of integrability of a Hamiltonian system: real integrability (given by differentiable or analytical real first integrals), algebraic integrability (the Liouville torus becomes the affine part of abelian varieties), integrability by algebraic first integrals, by holomorphic first integrals or by meromorphic first integrals.

A second elementary but important remark is that if we are studying some kind of integrability, the allowed changes of variables (i.e., canonical transformations) must belong to the same category. For instance, if we are studying the integrability of a system by rational first integrals (or if it has the Painlevé property), changes of variables must have the same degree of regularity. So, changes of variables that introduce essential singularities in the phase variables (respectively, in the time) are not allowed, because the above integrability is not preserved by these changes (respectively, the Painlevé property). By all the above, the Bianchi IX cosmological model is a good "laboratory" in order to understand the interplay between the several concepts of integrability.

In this section we shall prove the non-integrability of the Bianchi IX model (with the Hamiltonian given by H above) by rational first integrals.

5.2.2 Non-integrability

Our proof of non-integrability lies on the variational equations along a particular solution of the Taub family.

As noticed by Taub, the subspace $x_2 = x_3$ ($\dot{x}_2 = \dot{x}_3$) is invariant by the flow of the system, and the reduction to this invariant four dimensional space (three dimensional on the restricted manifold M_0) is completely integrable (it is an integrable subsystem) and their solutions can be calculated explicitly ([99], p. 481). From the Taub family of solutions we select the particular ones

$$x_1 = \frac{2k}{\cosh(2kt)}, \quad x_2 = x_3 = \frac{k\cosh(2kt)}{2\cosh^2(kt)},$$

k being a parameter. This particular integral curve is our integral curve Γ, along which we compute the VE. We remark that Γ is contained in the energy constraint M_0. Furthermore $\overline{\Gamma}$ is the Riemann sphere $\mathbf{P}^1$, since it has a global rational parametrization in the variable $z := \tanh kt$. Effectively, we can write $x_1, x_2 = x_3, \dot{x}_1, \dot{x}_2 = \dot{x}_3$ as rational functions of z,

$$x_1 = \frac{2k(1-z^2)}{1+z^2}, \qquad x_2 = x_3 = \frac{k(1+z^2)}{2},$$

$$\dot{x}_1 = -\frac{8k^2 z(1-z^2)}{(1+z^2)^2}, \quad \dot{x}_2 = \dot{x}_3 = kz(1-z^2).$$

The relevant part of the VE is given by the Normal Variational Equation (NVE) transversal to the invariant space $x_2 = x_3$, $\dot{x}_2 = \dot{x}_3$, obtained in [61] (as we know from Chapter 4, the tangential part of the VE is integrable)

$$\ddot{\alpha} - 2(x_1 x_2 - 2x_2^2)\alpha = 0. \tag{5.5}$$

Now, as noted in [61], it is possible to write the NVE over the Riemann sphere $\mathbf{P}^1$ (i.e., with rational coefficients) by the change $x := z^2 = \tanh^2(kt)$. We write the obtained equation in its invariant reduced form (i.e., without the term in the first derivative in the standard way) in the new independent variable η as

$$\frac{d^2\eta}{dx^2} + \left(\frac{1}{4}\frac{1}{x-1} + \frac{5}{4}\frac{1}{(x-1)^2} + \frac{3}{16}\frac{1}{x^2}\right)\eta = 0. \tag{5.6}$$

We remark that in the above equation the parameter k is missing.

The above equation has regular singularities at $x = 0, 1$ and an irregular singularity at $x = \infty$ (it is a confluent Heun's Equation). The dynamical meaning of the equilibrium points is the following. The natural parametrization of the Riemann surface $\overline{\Gamma} \approx \mathbf{P}^1$ is the variable z, with $x = z^2$. Then by the double covering $\mathbf{P}^1 \longrightarrow \mathbf{P}^1$, $z \mapsto x$, from the points $x = 0, 1, \infty$, we get the four points $z = 0, \pm 1, \infty$ of $\bar{\Gamma}$. It is a direct consequence of the rational parametrization of $\bar{\Gamma}$ in the parameter z, obtained above, that $z = 0, \pm 1$ (hence $x = 0, 1$) are equilibrium points, with $x_1(0) = 2k$, $x_2(0) = x_3(0) = k/2$, $\dot{x}_i(0) = 0$, $i = 1, 2, 3$ and $x_1(\pm 1) = 0$, $x_2(\pm 1) = x_3(\pm 1) = k$, $x_i(\pm 1) = 0$, $i = 1, 2, 3$, respectively. The point $x = \infty$ corresponds to the point $z = \infty$ of the phase curve $\bar{\Gamma}$.

We remark that the points $z = \pm 1$ (i.e., $x = 1$) correspond also to the physical gravitational collapse, because

$$x_1 x_2 x_3 = \frac{k^3}{2}(1 - z^4) = \frac{k^3}{2}(1 - x^2).$$

Now, we can apply the Kovacic Algorithm (Section 2.7).
Let

$$r(x) = -\left(\frac{1}{4}\frac{1}{x-1} + \frac{5}{4}\frac{1}{(x-1)^2} + \frac{3}{16}\frac{1}{x^2}\right) := \frac{s(x)}{t(x)},$$

with t a monic polynomial. Then the algorithm is divided into three steps:

First Step

1.1. The sets Γ and Γ' are

$$\Gamma = \{0, 1, \infty\}, \quad \Gamma' = \{0, 1\},$$

with orders $o(0) = o(1) = 2$, $o(\infty) = \max(0, 4 + \deg s - \deg t) = 3$, and we define $m^+ := \max(m, o(\infty)) = \max(2, o(\infty)) = 3$ (m is the cardinal of Γ').
Then the set of singular points is classified by the order:

$$\Gamma_2 = \{0, 1\}, \quad \Gamma_3 = \{\infty\}.$$

1.2. As $m^+ > 2$, we get $\gamma_2 = 2$, and $\gamma := \gamma_2 + \#\Gamma_3 = 3$.

1.3. $\alpha_0 = -\frac{3}{16}$, $\alpha_1 = -\frac{5}{4}$.

1.5. $L = \{2\}$, $n = 2$.

Second Step

2.4. As $n = 2 \geq 2$ and $h(2) = 4$, we obtain $E_0 = \mathbf{Z} \cap \{1, 2, 3\} = \{1, 2, 3\}$, $E_1 = \mathbf{Z} \cap \{2(1 - \sqrt{-4}), \quad 2, \quad 2(1 + \sqrt{-4})\} = \{2\}$.
2.6. $E_\infty = \{3\}$.

Third Step

3.1. If $e_1 = (1, 2, 3)$, $e_2 = (2, 2, 3)$, $e_3 = (3, 2, 3)$ (the elements of $E_0 \times E_1 \times E_\infty$), then the numbers $d(e_i)$, $i = 1, 2, 3$ are not natural numbers. Hence the Galois group of the equation (5.6) is $SL(2, \mathbf{C})$. This equation is non-integrable and the NVE is also non-integrable.

As the irregular point of the VE is at ∞ in the phase space, then by the suitable version of Corollary 4.5 (corresponding to Theorem 4.3), we have proved the following result.

Proposition 5.5 *The Bianchi IX Cosmological Model, considered as a Hamiltonian system given by the Hamilton function H above, is not completely integrable with rational first integrals.*

We note that although we have not worked in canonical coordinates (i.e., in a canonical frame in the VE) the above result is independent of this fact, because our formulation of the general theory in Chapter 4 was completely intrinsic.

5.3   Sitnikov's Three-Body Problem

5.3.1   The model

The Sitnikov system is a restricted three body problem given by a very symmetrical configuration: the primaries with equal masses m move in ellipses of eccentricity e in the XY plane around their center of masses O, and the third infinitesimal body moves along the axis OZ perpendicular to the plane where the primaries move [83]. We take, as usual, the normalization of units in such a way that $m = 1/2$, the period of the primaries is 2π and the gravitational constant is equal to one.

Then the equation of motion of the third body is given by

$$\ddot{z} + \frac{z}{(z^2 + r(t)^2)^{3/2}} = 0, \tag{5.7}$$

$r(t)$ being the distance of one of the primaries to the center of masses O.

In his book [83], Moser showed the non-integrability (and in fact the chaotic behaviour of this system) by proving that it contains a Bernoulli shift as a subsystem (in the real domain of the phase space). Here we will give an alternative proof of the analytical non-integrability of this system in the complex domain.

We follow [70] and we choose as a new time the eccentric anomaly τ. The transformation is given by the Kepler equation

$$t = \tau - e \sin \tau.$$

Then equation (5.7) is transformed to

$$\frac{dz}{d\tau} = (1 - e \cos \tau)v, \quad \frac{dv}{d\tau} = -\frac{(1 - e \cos \tau)z}{(z^2 + r(\tau)^2)^{3/2}}, \quad \text{with} \quad r(\tau) = \frac{1}{2}(1 - e \cos \tau).$$

5.3.2   Non-integrability

As the particular integral curve Γ we take the triple collision orbit with $e = 1$, $r = 1/2(1 - \cos\tau)$, $z = v = 0$. The NVE along Γ is given by

$$\frac{d^2\xi}{d\tau^2} - \frac{\sin\tau}{1 - \cos\tau}\frac{d\xi}{d\tau} + \frac{8}{1 - \cos\tau}\xi = 0. \tag{5.8}$$

Now, in order to get an equation over the Riemann sphere, we make the transformation

$$x = \frac{\cos\tau}{2} + 1.$$

We obtain

$$\frac{d^2\xi}{dx^2} + \left(\frac{1/2}{x} - \frac{1/2}{x-1}\right)\frac{d\xi}{dx} + 4\left(\frac{1}{(x-1)^2} - \frac{1}{x-1} + \frac{1}{x}\right)\xi = 0. \tag{5.9}$$

The above equation is a Riemann (or generalized hypergeometric) equation with three regular singular singularities at $x = 0$, $x = 1$ and $x = \infty$. The triple collision corresponds to the singular point $x = 1$ and the point $x = 0$ corresponds to a branching point of the double covering defined by the above change of variables $x \mapsto \tau$ (physically these two points correspond to the two vertexes of the degenerated ellipse with $e = 1$). We remark that by Theorem 2.5, the identity component of the Galois group of equation (5.8) is the same as the identity component of the Galois Group of equation (5.9).

For equation (5.9), the difference of exponents at $x = 0$, $x = 1$, $x = \infty$ is (respectively) $\lambda = 1/2$, $\mu = i/2\sqrt{55}$ and $\nu = 1$.

It is very easy to check condition (i) or (ii) of Kimura's theorem (Theorem 2.6) for equation (5.9): as the Galois group is not finite (the exponents at the point $x = 1$ are not rational numbers) we only need to check (i) and (ii), family 1. In a direct way we get the non-solvability of the identity component of the Galois group. In particular, this identity component is not abelian. Hence, we have obtained the following result.

Proposition 5.6 *In a neighborhood of the triple collision orbit, the Sitnikov Three Body Problem is not completely integrable with meromorphic first integrals.*

The author is indebted to José Martinez-Alfaro for suggesting this example.

Chapter 6

An Application of the Lamé Equation

Very simple examples of Hamiltonian systems in two degrees of freedom lead to *NVE* of Lamé type, equation (2.16),

$$\frac{d^2\xi}{dt^2} = \big(A\mathcal{P}(t) + B\big)\xi, \tag{6.1}$$

where $\mathcal{P}$ denotes the Weierstrass function and A and B are, in general, complex parameters. It is assumed, in what follows, that the roots of the polynomial f associated to $\mathcal{P}$ are simple (otherwise $\mathcal{P}$ is reduced to elementary functions). This is ensured if the discriminant

$$\Delta := 27g_3^2 - g_2^3 \tag{6.2}$$

is non-zero, where g_2 and g_3 are the associated invariants (see Chapter 2).

It was observed in the early examples [48], that a necessary condition for integrability was n integer, where we set (as in Chapter 2) $A = n(n+1)$. The motivation of this chapter is to understand this behaviour.

In the forthcoming sections we obtain, first, the potentials of classical Hamiltonians with an invariant plane such that the *NVE* are of Lamé type. Then a non-integrability criterion is obtained for these Hamiltonians. The results given here are not complete because we have not been able to prove that some numerical coefficients are different from zero (although a large number of them has been checked!). We conjecture that all of them are different from zero.

The chapter ends with the study of some old and new examples. The case of the homogeneous Hénon-Heiles potential is studied in detail, in particular, we give another proof of Proposition 5.4 and numerical evidence of the non-integrability of the system for $e = 2$.

The results of this chapter were obtained in a joint work of the author with C. Simó [82].

J. J. Morales Ruiz, *Differential Galois Theory and Non-Integrability of Hamiltonian*, Systems, Modern Birkhäuser Classics, DOI: 10.1007/978-3-0348-0723-4_6, © Springer Basel 1999

6.1   Computation of the potentials

Let

$$H = \frac{1}{2}\left(y_1^2 + y_1^2\right) + V(x_1, x_2)$$

be a two degrees of freedom classical Hamiltonian, where V is a real analytic function on some domain which will be considered in $\mathbf{C}^2$. Assume that there exists a continuous family of integral curves, Γ_h, parametrized by the energy, h, lying on an invariant plane that, for concreteness, will be taken as

$$\Gamma_h : x_2 = y_2 = 0, \quad x_1 = x_1(t, h), \quad y_1 = y_1(t, h).$$

We know from Chapter 4 that a necessary and sufficient condition is that

$$V(x_1, x_2) = \varphi(x_1) - \frac{1}{2}\alpha(x_1)\,x_2^2 + O(x_2^3), \tag{6.3}$$

where φ and α are arbitrary functions.
The *NVE* along Γ_h is

$$\ddot{\xi} - \alpha(t, h)\,\xi = 0, \tag{6.4}$$

where, for simplicity, we denote by $\alpha(t, h)$ what in fact is $\alpha(x_1(t, h))$, and $x_1(t, h)$ is a solution of

$$\frac{1}{2}\dot{x}_1^2 + \varphi(x_1) = h,$$

the energy h ranging in a real interval. We want to obtain a potential, V, of the type (6.3) (that is, to obtain the functions φ and α, the $O(x_2^3)$ being arbitrary) such that (6.4) is of the type (6.1), that is

$$\alpha(t, h) = A(h)\,\mathcal{P}(t, h) + B(h), \tag{6.5}$$

A and B being parameters and $\mathcal{P}$ the Weierstrass elliptic function. From now on we keep in mind that everything depends on h, but we do not write it explicitly.

From (6.4) and using $'$ to denote $\frac{d}{dx_1}$, it follows

$$\dot{\alpha}^2(t) = 2\,\alpha'^2(x_1)\,h - 2\,\alpha'^2(x_1)\,\varphi(x_1). \tag{6.6}$$

Assume $\alpha(x_1)$ not identically constant. Hence, we can obtain $x_1 = x_1(\alpha)$ (possibly multivaluated). Hence, as from (6.5) and from the differential equation which satisfies $\mathcal{P}(t)$ ($\frac{d\mathcal{P}}{dt} = f(\mathcal{P})$), it follows that $\dot{\alpha}^2$ is a cubic polynomial in α, depending also on h. By comparing with (6.6) we get

$$\dot{\alpha}^2 = P(\alpha, h) = P_1(\alpha) + h\,P_2(\alpha), \tag{6.7}$$

where P is a polynomial of degree 3 in α and, therefore, either P_1 or P_2 must have degree 3.

We remark that the case $\alpha = \text{constant} = B$ gives a separable potential up to the $O(x_2^3)$ terms. This is equivalent to $P_2 \equiv 0$ (see (6.8) below).

Hence, by comparing (6.6) and (6.7) and denoting by $\varphi(\alpha)$ the function $\varphi(x_1(\alpha))$ we have

$$\varphi(\alpha) = -\frac{P_1(\alpha)}{P_2(\alpha)} \ , \quad \alpha'^2(x_1) = \frac{1}{2} P_2(\alpha(x_1)) \ . \tag{6.8}$$

From (6.8) we obtain the potential from P_1 and P_2 by using the scheme

$$\begin{array}{ccccc} P_2(\alpha) & \rightarrow & \alpha(x_1) & \rightarrow & \varphi(x_1) \\ P_1(\alpha) & \rightarrow & \varphi(\alpha) & & \end{array} . \tag{6.9}$$

Let g_2 and g_3 be the invariants of $\mathcal{P}$. Now we look for a relation between h, P_1 and P_2, on one side, and A, B, g_2 and g_3, on the other. From (6.5) we obtain

$$u := \mathcal{P}(t) = \frac{1}{A}\left(\alpha(t) - B\right) \ , \quad v := \dot{\mathcal{P}}(t) = \frac{\dot\alpha(t)}{A} \ , \quad \dot\alpha^2(t) = A^2 v^2 = A^2\left(4u^3 - g_2 u - g_3\right).$$

Therefore,

$$P(\alpha, h) = \dot\alpha^2(t) = \frac{4}{A}\alpha^3 - \frac{12B}{A}\alpha^2 + \left(\frac{12B^2}{A} - g_2 A\right)\alpha - \frac{4B^3}{A} + g_2 AB - g_3 A^2 \ . \tag{6.10}$$

We introduce the coefficients $a_1, \ldots, d_2$ by setting

$$P(\alpha, h) = \left(a_1 + h\, a_2\right)\alpha^3 + \left(b_1 + h\, b_2\right)\alpha^2 + \left(c_1 + h\, c_2\right)\alpha + d_1 + h\, d_2 \ . \tag{6.11}$$

By comparing (6.10) and (6.11) we obtain

$$\frac{4}{A} = a_1 + h\, a_2, \tag{6.12}$$

$$-\frac{12B}{A} = b_1 + h\, b_2, \tag{6.13}$$

$$\frac{12B^2}{A} - g_2 A = c_1 + h\, c_2, \tag{6.14}$$

$$-\frac{4B^3}{A} + g_2 AB - g_3 A^2 = d_1 + h\, d_2. \tag{6.15}$$

Let us proceed to the effective computation of the potentials. We classify them according to the degree of P_2 and then we use (6.8). In the expressions below e denotes an integration constant. We restrict ourselves to real solutions.

(A) $\deg P_2 = 0 \Rightarrow P_2(\alpha) = d_2 > 0$, $\alpha = \pm\sqrt{\frac{d_2}{2}}\,x_1 + e$.

(B) $\deg P_2 = 1 \Rightarrow P_2(\alpha) = c_2\alpha + d_2$, $\alpha = \frac{c_2}{8}x_1^2 + e\,x_1 + \frac{2e^2 - d_2}{c_2}$.

(C) $\deg P_2 = 2 \Rightarrow P_2(\alpha) = b_2\,\alpha^2 + c_2\,\alpha + d_2$. Let $D := c_2^2 - 4b_2\,d_2$.

We should consider three cases:

(C.1) $D = 0$, $b_2 > 0$: $P_2(\alpha) = b_2\left(\alpha + \frac{c_2}{2b_2}\right)^2$, $\alpha = e \cdot \exp\left(\left(\frac{b_2}{2}\right)^{1/2}x_1\right) - \frac{c_2}{2b_2}$.

(C.2) $D < 0$, $b_2 > 0$: $\alpha = \frac{\sqrt{-D}}{2b_2}\sinh\left(\left(\frac{b_2}{2}\right)^{1/2}x_1 + e\right) - \frac{c_2}{2b_2}$.

(C.3.1) $D > 0$, $b_2 > 0$, $\alpha = \frac{\sqrt{D}}{2b_2}\cosh\left(\left(\frac{b_2}{2}\right)^{1/2}x_1 + e\right) - \frac{c_2}{2b_2}$.

(C.3.2) $D > 0$, $b_2 < 0$, $\alpha = \frac{\sqrt{D}}{2b_2}\sin\left(\left(-\frac{b_2}{2}\right)^{1/2}x_1 + e\right) - \frac{c_2}{2b_2}$. (D) $\deg P_2 = 3$.

From (6.7) one derives

$$\alpha(x_1) = C\,\beta(x_1) + E, \quad \text{with} \quad \beta'^2(x_1) = 4\beta^3 - \bar{g}_2\,\beta - \bar{g}_3,$$

$$\text{where} \quad a_2 = \frac{8}{C}, \quad b_2 = -\frac{24E}{C}, \quad c_2 = -2C\,\bar{g}_2 + \frac{24E^2}{C},$$

$$d_2 = -2C^2\,\bar{g}_3 + 2CE\,\bar{g}_2 - 8\frac{E^3}{C}.$$

Let $\bar{\Delta} = 27\bar{g}_3^2 - \bar{g}_2^3$. There are three possibilities:

(D.1) $\bar{\Delta} \neq 0$, and then $\beta = \mathcal{P}(x_1 + e)$.

(D.2) $\bar{\Delta} = 0$ and two of the roots of $4\beta^3 - \bar{g}_2\,\beta - \bar{g}_3$ are equal. There are two subcases (see, for instance, [9], Vol. I p. 27). Let $\bar{e}_3 \le \bar{e}_2 \le \bar{e}_1$ be the roots.

(D.2.1) $\bar{e}_2 = \bar{e}_3 = -\frac{1}{2}\bar{e}_1 < \bar{e}_1$. Then

$$\beta = -\frac{3}{2}\frac{\bar{g}_3}{\bar{g}_2} + \frac{9}{2}\frac{\bar{g}_3}{\bar{g}_2}\,\mathrm{cosec}^2\left(\left(\frac{9\bar{g}_3}{2\bar{g}_2}\right)^{1/2}x_1 + e\right).$$

(D.2.2) $\bar{e}_1 = \bar{e}_2 = -\frac{1}{2}\bar{e}_3 > \bar{e}_3$. Then

$$\beta = \frac{3}{2}\frac{\bar{g}_3}{\bar{g}_2} - \frac{9}{2}\frac{\bar{g}_3}{\bar{g}_2}\,\coth^2\left(\left(-\frac{9\bar{g}_3}{2\bar{g}_2}\right)^{1/2}x_1 + e\right).$$

(D.3) $\bar{\Delta} = 0$ and the three roots are equal. We can write $P_2(\alpha) = a_2(\alpha - \bar{e}_1)^3$ and, hence,

$$\alpha = \frac{8}{a_2(x_1 + e)^2} + \bar{e}_1.$$

In all the cases, when α is available as a function of x_1, the function φ is obtained from (6.8). We refer to Section 6.3, where some realizations of the above cases are given explicitly.

6.2 Non-integrability criterion

Given a two degrees of freedom complex analytic Hamiltonian system having an integral curve, Γ, we have

Theorem 6.1 ([75, 81]) *If the* NVE *along* Γ *is of Lamé type, (i.e., (6.1)), in the temporal parametrization and falls outside of the solvability cases (i), (ii), (iii) of Proposition 2.6, then the Hamiltonian system is non-integrable.*

Proof. The proof follows in a direct way from Proposition 2.6 and our general theorems of Section 4.2 (Corollary 4.5), because if we are not in the cases (i), (ii) or (iii), the identity component of the Galois group is not solvable and, in particular, it is not abelian. $\square$

In this chapter we fix the kind of regularity of the first integrals: when there is no global meromorphic first integral independent of the Hamiltonian, we say that the system is non-integrable.

Next we study the non-integrable cases of the families of potentials A), B), C) and D) obtained in Section 6.1, using the above theorem. We notice that, for a given potential (that is, given the polynomials P_1 and P_2 of Section 6.1) we have a one parameter family of Lamé equations, the parameter being the level of energy, h. The analysis of the D) families, which have $a_2 \neq 0$, is elementary. It is enough to consider that, according to (6.12) we reach irrational values of n when h changes. Hence, all the D) families are non-integrable.

Therefore we can assume $a_2 = 0$ (families A), B), C)) so that for a given potential the value of n remains fixed when h changes. In particular one can not jump from one of the cases (i), (ii), (iii) (of Proposition 2.6) to another when h changes.

If our system falls in case (i) we can not derive additional integrability conditions from the analysis of the *NVE*. Before proceeding to the case (ii) we analyze the case (iii). Hence, assume that for a given value of h we have a Lamé equation of type (iii). We have $a_1 = \frac{4}{n(n+1)}$, with n as in Subsection 2.8.4 (iii).

Lemma 6.1 *Consider the curve* $\sigma : h \rightarrow \big(j(h), B(h)\big)$ *defined by means of* (6.12)–(6.15) *with* $a_2 = 0$ *and the formula (2.18) of Chapter 2. Then* σ *changes continuously with respect to* h *except in the cases:*
1) $P_2 \equiv 0$, *which, by a remark above, has* $n = 0$, *hence it is not really of the Lamé type,*
2) $b_2 = 0$, $c_2 = 0$, $b_1^2 - 3c_1a_1 = 0$,
3) $b_2 = 0$, $c_2b_1 - 3a_1d_2 = 0$, $2b_1^3 - 9a_1b_1c_1 + 27a_1^2d_1 = 0$.

Proof. From (6.13) we should have $b_2 = 0$ (otherwise B changes linearly with h). The possibilities to have $j(h)$ constant are $g_2(h) \equiv 0$, $g_3(h) \equiv 0$ (both cases can not occur simultaneously because the discriminant, Δ, must satisfy $\Delta \not\equiv 0$) or both g_2 and g_3 are independent of h.

The condition $g_2 \equiv 0$ gives 2) and $g_3 \equiv 0$ gives 3). If g_2 and g_3 do not depend on h one finds $c_2 = d_2 = 0$. But as $a_2 = b_2 = 0$, we have $P_2 \equiv 0$. $\quad\square$

From the Dwork result (Proposition 2.8) and from Theorem 6.1 follows immediately the next result.

Proposition 6.1 *If $a_2 = 0$, the NVE are of Lamé type, not of the types (i) or (ii), and the Hamiltonian system is integrable, then one should have $b_2 = 0$ and either $c_2 = 0$, $b_1^2 - 3c_1a_1 = 0$ or $c_2b_1 - 3a_1d_2 = 0$, $2b_1^3 - 9a_1b_1c_1 + 27a_1^2d_1 = 0$.*

Now we start the discussion of case (ii) (of Proposition 2.6). We need to make an elementary but detailed analysis of the algebraic structure of the Brioschi determinant in the framework of this chapter.

To have integrability, a necessary condition is $Q_m\left(\frac{g_2}{4}, \frac{g_3}{4}, B\right) \equiv 0$ as a polynomial in h, provided $\Delta \neq 0$. We want to express the conditions for integrability in terms of the coefficients $a_i, b_i, c_i, d_i, i = 1, 2$ of the polynomials appearing in the potential. We recall that in case (ii) one should have $a_1 = \frac{16}{4m^2-1}$ for some $m \in \mathbf{N}$, $a_2 = 0$, the remaining coefficients being arbitrary, except that b_2, c_2 and d_2 can not be zero simultaneously.

Let us introduce $\bar{B} = \bar{b}_1 + h\bar{b}_2 \equiv b_1 + h b_2$, $\bar{C} = \bar{c}_1 + h\bar{c}_2 \equiv (c_1 + h c_2)\frac{a_1}{16}$, $\bar{D} = \bar{d}_1 + h\bar{d}_2 \equiv (d_1 + h d_2)\frac{a_1^2}{64}$.
Then $B = -\frac{\bar{B}}{3a_1}$, $\frac{g_2}{4} = \frac{\bar{B}^2}{48} - \bar{C}$, $\frac{g_3}{4} = -\frac{\bar{B}^3}{864} + \frac{\bar{B}\bar{C}}{12} - \bar{D}$ and the discriminant has the expression

$$\Delta = \bar{B}^3\bar{D} - \bar{B}^2\bar{C}^2 - 72\bar{B}\bar{C}\bar{D} + 64\bar{C}^3 + 432\bar{D}^2.$$

Multiplying each column in the Brioschi determinant $Q_m(g_2/4, g_3/4, B)$ (see Section 2.8) by 864 we get the $m \times m$ determinant $\mathcal{D}_m(\bar{B}, \bar{C}, \bar{D}) :=$

$$\begin{vmatrix} -18(4m^2-1)\bar{B} & 864(m-1) \\ (2m-1)(m-1)W_1 & -18(4m^2-1)\bar{B} & 864\cdot 2(m-2) \\ (2m-1)(2m-2)W_2 & (2m-2)(m-2)W_1 & -18(4m^2-1)\bar{B} & 864\cdot 3(m-3) \\ & (2m-2)(2m-3)W_2 & (2m-3)(m-3)W_1 & -18(4m^2-1)\bar{B} & \ddots \\ & & (2m-3)(2m-4)W_2 & (2m-4)(m-4)W_1 & \ddots \\ & & & (2m-4)(2m-5)W_2 & \ddots \\ & & & & \ddots \end{vmatrix},$$

where W_1 stands for $18\bar{B}^2 - 864\bar{C}$ and W_2 for $-\bar{B}^3 + 72\bar{B}\bar{C} - 864\bar{D}$. It follows immediately that $\mathcal{D}_m(\bar{B}, \bar{C}, \bar{D}) = \sum_{i+2j+3k=m} C_{i,j,k}\bar{B}^i\bar{C}^j\bar{D}^k$, where $C_{i,j,k}$ are integer coefficients.

Proposition 6.2 *Provided some constants are different from zero (see the proof for the details) necessary conditions for integrability in case (ii), assuming $\Delta \neq 0$, are:*

1. *If $m \equiv 1 \pmod{6}$: $\bar{B} \equiv 0$, $\bar{C} \equiv 0$ or $\bar{B} \equiv 0, \bar{D} \equiv 0$,*
2. *If $m \equiv 2 \pmod{6}$: $\bar{B} \equiv 0$, $\bar{C} \equiv 0$,*
3. *If $m \equiv 3 \pmod{6}$: $\bar{B} \equiv 0$, $\bar{D} \equiv 0$,*
4. *If $m \equiv 4 \pmod{6}$: $\bar{B} \equiv 0$, $\bar{C} \equiv 0$,*
5. *If $m \equiv 5 \pmod{6}$: $\bar{B} \equiv 0$, $\bar{C} \equiv 0$ or $\bar{B} \equiv 0, \bar{D} \equiv 0$,*

the case $m \equiv 0 \pmod{6}$ being always non-integrable.
In the list above the following exceptions occur:

1. *If $m = 1$ the condition is only $\bar{B} \equiv 0$,*
2. *If $m = 2$ it is $\bar{b}_2 = \bar{c}_2 = 0$, $\bar{c}_1 = -3\bar{b}_1^2/256$,*
3. *If $m = 3$ we need $\bar{b}_2 = 0$, $\bar{d}_2 = -11\bar{b}_1\bar{c}_2/64$ and $\bar{d}_1 = -11\bar{b}_1\bar{c}_1/64 - 45\bar{d}_1^3/65536$.*

Proof. For given values of $m, \bar{b}_1, \bar{b}_2, \bar{c}_1, \bar{c}_2, \bar{d}_1$ and $\bar{d}_2$ we shall consider $\mathcal{D}_m(\bar{B}, \bar{C}, \bar{D})$ as a function of h. It must be identically zero. This will impose conditions on the coefficients above.

If $m = 1$ then $\mathcal{D}_m = \bar{B}$ and hence $\bar{b}_1 = \bar{b}_2 = 0$.

If $m = 2$ then $\mathcal{D}_m = 8748(3\bar{B}^2 + 256\bar{C})$. The term in h^2 contains $\bar{b}_2^2$ and non-null factors. Hence $\bar{b}_2 = 0$. Then the term in h contains $\bar{c}_2$ and non-null factors. Also $\bar{c}_2$ must be zero. Finally, the independent term $3\bar{b}_1^2 + 256\bar{c}_1$ must be zero and this case is ended.

If $m = 3$ then $\mathcal{D}_m = C_{3,0,0}\bar{B}^3 + C_{1,1,0}\bar{B}\bar{C} + C_{0,0,3}\bar{D}$, where $C_{3,0,0} = -35429400$ and $C_{1,1,0} = -8868372480$, $C_{0,0,1} = -51597803520$. The term in h^3 is $C_{3,0,0}\bar{b}_2^3$ and, hence, $\bar{b}_2 = 0$. Then the terms in h^1, h^0 give the other two conditions.

Now we proceed to the general cases according to the class of m modulo 6.

If $m \equiv 0 \pmod{6}$ the highest power of h appears in $C_{m,0,0}\bar{B}^m$. Assuming $C_{m,0,0} \neq 0$ we should have $\bar{b}_2 = 0$. We note here that this is a general fact, independent of the value of m, provided $C_{m,0,0} \neq 0$. Then the highest power of h appears in $C_{0,\frac{m}{2},0}\bar{C}^{\frac{m}{2}}$. Again we must have $\bar{c}_2 = 0$ if $C_{0,\frac{m}{2},0} \neq 0$. But then the highest power of h appears in $C_{0,0,\frac{m}{3}}\bar{D}^{\frac{m}{3}}$, and if $C_{0,0,\frac{m}{3}} \neq 0$ we must have $\bar{d}_2 = 0$, but as $\bar{b}_2^2 + \bar{c}_2^2 + \bar{d}_2^2 > 0$ the system is non-integrable.

If $m \equiv 2 \pmod{6}$ and $C_{m,0,0} \neq 0$, $C_{0,\frac{m}{2},0} \neq 0$ we have $\bar{b}_2 = \bar{c}_2 = 0$. Now the dominant terms in h come from $\bar{B}^2\bar{D}^{\frac{m-2}{3}}$ and $\bar{C}\bar{D}^{\frac{m-2}{3}}$. As $\bar{d}_2 \neq 0$ we must have $C_{2,0,\frac{m-2}{3}}\bar{b}_1^2 + C_{0,1,\frac{m-2}{3}}\bar{c}_1 = 0$. The next highest power of h appears in the $\bar{B}^5\bar{D}^{\frac{m-5}{3}}$, $\bar{B}^3\bar{C}\bar{D}^{\frac{m-5}{3}}$ and $\bar{B}\bar{C}^2\bar{D}^{\frac{m-5}{3}}$ terms. As a factor of $(\bar{d}_2h)^{\frac{m-5}{3}}$ we have

$$C_{5,0,\frac{m-5}{3}}\bar{b}_1^5 + C_{3,1,\frac{m-5}{3}}\bar{b}_1^3\bar{c}_1 + C_{1,2,\frac{m-5}{3}}\bar{b}_1\bar{c}_1^2, \tag{6.16}$$

which must be zero. If $C_{0,1,\frac{m-2}{3}} \neq 0$ we obtain $\bar{c}_1$ in terms of $\bar{b}_1$ and, inserting in (6.16) we get

$$\bar{b}_1^5 \left[C_{5,0,\frac{m-5}{3}} - C_{3,1,\frac{m-5}{3}} \frac{C_{2,0,\frac{m-2}{3}}}{C_{0,1,\frac{m-2}{3}}} + C_{1,2,\frac{m-5}{3}} \left(\frac{C_{2,0,\frac{m-2}{3}}}{C_{0,1,\frac{m-2}{3}}} \right)^2 \right] = 0 . \qquad (6.17)$$

If the term inside square brackets in (6.17) is different from zero we should have $\bar{b}_1 = 0$, and then $\bar{c}_1 = 0$. Summarizing, we should have $\bar{B} \equiv \bar{C} \equiv 0$ but $\bar{D}$ is arbitrary, ending this case.

If $m \equiv 4 \pmod 6$ proceeding as in the previous case we have $\bar{b}_2 = \bar{c}_2 = 0$ if $C_{m,0,0} \neq 0$, $C_{0,\frac{m}{2},0} \neq 0$. Then the dominant term appears in $\bar{B}\bar{D}^{\frac{m-1}{3}}$ and, as $\bar{d}_2 \neq 0$, we must have $\bar{b}_1 = 0$ provided $C_{1,0,\frac{m-1}{3}} \neq 0$. The next dominant term appears in $\bar{C}^2\bar{D}^{\frac{m-4}{3}}$. Again if $C_{0,2,\frac{m-4}{3}} \neq 0$ we must have $\bar{c}_1 = 0$, ending the proof in this case. Hence $\bar{B} \equiv 0$, $\bar{C} \equiv 0$, $\bar{D}$ being arbitrary.

We proceed to the cases with m odd. As we shall see, a part of the proof is common for the three cases. We start with the non-common part. We assume $C_{m,0,0} \neq 0$ and hence $\bar{b}_2 = 0$ in all cases.

If $m \equiv 1 \pmod 6$, $m > 1$, the dominant terms appear in $\bar{B}\bar{C}^{\frac{m-1}{2}}$ and $\bar{C}^{\frac{m-3}{2}}\bar{D}$, and the coefficient of $h^{\frac{m-1}{2}}$ is

$$\bar{c}_2^{\frac{m-3}{2}} \left(C_{1,\frac{m-1}{2},0}\bar{b}_1\bar{c}_2 + C_{0,\frac{m-3}{2},1}\bar{d}_2 \right). \qquad (6.18)$$

If $\bar{c}_2 = 0$ then the dominant term is $\bar{B}\bar{D}^{\frac{m-1}{3}}$ and, as $\bar{d}_2 \neq 0$, we must have $\bar{b}_1 = 0$ provided $C_{1,0,\frac{m-1}{3}} \neq 0$. But then, if $C_{0,2,\frac{m-4}{3}} \neq 0$ the dominant power of h appears in $\bar{C}^2\bar{D}^{\frac{m-4}{3}}$ and we must have $\bar{c}_1 = 0$. Hence one possibility is $\bar{B} \equiv 0$, $\bar{C} \equiv 0$.

If $\bar{c}_2 \neq 0$ the second factor in (6.18) must be zero. Assume $\bar{b}_1 = 0$ and $C_{0,\frac{m-3}{2},1} \neq 0$. Then we must have $\bar{d}_2 = 0$. The current dominant term is now $C_{0,\frac{m-3}{2},1}\bar{c}_2^{\frac{m-3}{2}} \bar{d}_1 h^{\frac{m-3}{2}}$, and we must have $\bar{d}_1 = 0$. Therefore another possibility is $\bar{B} \equiv 0$, $\bar{D} \equiv 0$.

It remains to discuss the case $\bar{c}_2 \neq 0$, $C_{1,\frac{m-1}{2},0}\bar{b}_1\bar{c}_2 + C_{0,\frac{m-3}{2},1}\bar{d}_2 = 0$, $\bar{b}_1 \neq 0$, which we postpone for a joint discussion with the other m odd cases.

If $m \equiv 3 \pmod 6$, proceeding as before, $\bar{b}_2 = 0$ and either $\bar{c}_2 = 0$ or $C_{1,\frac{m-1}{2},0}\bar{b}_1\bar{c}_2 + C_{0,\frac{m-3}{2},1}\bar{d}_2 = 0$. If we assume $\bar{c}_2 = 0$ then the dominant term is $C_{0,0,\frac{m}{3}} \bar{d}_2^{\frac{m}{3}} h^{\frac{m}{3}}$, provided $C_{0,0,\frac{m}{3}} \neq 0$. But as $\bar{b}_2^2 + \bar{c}_2^2 + \bar{d}_2^2 \neq 0$ this case must be discarded. Hence it is the second term that must be zero, and proceeding as in the $m \equiv 1 \pmod 6$ case, if $\bar{b}_1 = 0$ and $C_{0,\frac{m-3}{2},1} \neq 0$ we must have $\bar{d}_2 = 0$, $\bar{d}_1 = 0$, i.e., we have $\bar{B} \equiv 0$, $\bar{D} \equiv 0$.

If $m \equiv 5 \pmod 6$ we must have $\bar{b}_2 = 0$ and either $\bar{c}_2 = 0$ or $C_{1,\frac{m-1}{2},0}\bar{b}_1\bar{c}_2 + C_{0,\frac{m-3}{2},1}\bar{d}_2 = 0$. If we assume $\bar{c}_2 = 0$, the dominant power of h has the coefficient

$$\left(C_{2,0,\frac{m-2}{3}}\bar{b}_1^2 + C_{0,1,\frac{m-2}{3}}\bar{c}_1\right)\bar{d}_1^{\frac{m-2}{3}}. \tag{6.19}$$

As $\bar{d}_2 \neq 0$ the coefficient in (6.19) must be zero. The next contribution appears in $h^{\frac{m-5}{3}}$, having as coefficient

$$\left(C_{5,0,\frac{m-5}{3}}\bar{b}_1^5 + C_{3,1,\frac{m-5}{3}}\bar{b}_1^3\bar{c}_1 + C_{1,2,\frac{m-5}{3}}\bar{b}_1\bar{c}_1^2\right)\bar{d}_2^{\frac{m-5}{3}}, \tag{6.20}$$

and the coefficient in (6.20) must also be zero. Now we proceed as in the case $m \equiv 2 \pmod 6$ and, under the same assumptions on the numerical coefficients, we have $\bar{b}_1 = 0$, $\bar{c}_1 = 0$. This gives the $\bar{B} \equiv 0$, $\bar{C} \equiv 0$ case.

If $\bar{c}_2 \neq 0$ we proceed as in the $m \equiv 1 \pmod 6$ case. If $\bar{b}_1 = 0$ we also proceed as in the $m \equiv 1 \pmod 6$ case.

It remains to study the odd m cases assuming $\bar{b}_2 = 0$, $\bar{c}_2 \neq 0$, $\bar{b}_1 \neq 0$. This requires $C_{m,0,0} \neq 0, C_{0,\frac{m-3}{2},1} \neq 0$. We look at the coefficients of $h^{\frac{m-1}{2}}, h^{\frac{m-3}{2}}, h^{\frac{m-5}{2}}$. Taking $\bar{C}^{\frac{m-15}{2}}$ as a factor (eventually the exponent can be negative) we should look for the coefficients of h^7, h^6 and h^5 in

$$\sum_{k=0}^{1} C_{1-k,\frac{m-1}{2}-k,k}\bar{B}^{1-k}\bar{C}^{7-k}\bar{D}^k + \sum_{k=0}^{3} C_{3-k,\frac{m-3}{2}-k,k}\bar{B}^{3-k}\bar{C}^{6-k}\bar{D}^k$$
$$+ \sum_{k=0}^{5} C_{5-k,\frac{m-5}{2}-k,k}\bar{B}^{5-k}\bar{C}^{5-k}\bar{D}^k. \tag{6.21}$$

As $\bar{c}_2 \neq 0, \bar{b}_1 \neq 0, \bar{b}_2 = 0$ we can use $\bar{E} := \bar{B}\bar{C}$ as independent variable instead of h. Then $\bar{D}$ can be written as $\mu_1 + \mu_2\bar{E}$, where μ_1, μ_2 are suitable numerical coefficients.

Let

$$P_1(\bar{E}) = \sum_{k=0}^{1} C_{1-k,\frac{m-1}{2}-k,k}\,\bar{E}^{1-k}\left(\mu_1 + \mu_2\bar{E}\right)^k,$$

$$P_3(\bar{E}) = \sum_{k=0}^{3} C_{3-k,\frac{m-3}{2}-k,k}\,\bar{E}^{3-k}\left(\mu_1 + \mu_2\bar{E}\right)^k,$$

$$P_5(\bar{E}) = \sum_{k=0}^{5} C_{5-k,\frac{m-5}{2}-k,k}\,\bar{E}^{5-k}\left(\mu_1 + \mu_2\bar{E}\right)^k.$$

Then (6.21) can be written as

$$\frac{\bar{E}^6}{\bar{B}^6}\,P_1(\bar{E}) + \frac{\bar{E}^3}{\bar{B}^3}\,P_3(\bar{E}) + P_5(\bar{E}). \tag{6.22}$$

The condition that the terms in $\bar{E}^7$, $\bar{E}^6$, $\bar{E}^5$ in (6.21) should be zero gives

$$P_1'(0) = 0\,,\ \ P_1(0) + \frac{\bar{B}^3}{3!}\,P_3'''(0) = 0\,,\ \ \frac{1}{2!}\,P_3''(0) + \frac{\bar{B}^3}{5!}\,P_5^V(0) = 0\,,$$

or, more explicitly,

$$C_{1,\frac{m-1}{2},0} + C_{0,\frac{m-3}{2},1}\,\mu_2 = 0, \tag{6.23}$$

$$\mu_1 C_{0,\frac{m-3}{2},1} + \bar{B}^3\,P_3(\mu_2) = 0\,,\ \ \mu_1\,P_3'(\mu_2) + \bar{B}^3\,P_5(\mu_2) = 0\,. \tag{6.24}$$

From (6.24) we derive $C_{0,\frac{m-3}{2},1}\,P_5(\mu_2) - P_3(\mu_2)\,P_3'(\mu_2) = 0$, where μ_2 is obtained from (6.23). If this condition is not satisfied then one should have $b_1 = 0$ and, therefore, the case $b_1 \neq 0$ must be discarded.

This ends the proof of Proposition 6.2 provided some numerical coefficients are non-zero. We proceed to prove this for some of them.

The coefficient $C_{m,0,0}$ is the value of $\mathcal{D}_m(\bar{B},\bar{C},\bar{D})$ when we set $\bar{B} = 1$, $\bar{C} = \bar{D} = 0$. Let $\Delta_{m,k}$ be the determinant obtained when in $\mathcal{D}_m$ we consider the first k rows and columns. Let $\bar{\Delta}_{m,k} = \Delta_{m,k}/18^k$. Then one has the following recurrence for $\bar{\Delta}_{m,k}$:

$$\bar{\Delta}_{m,k+1} = -(4m^2 - 1)\bar{\Delta}_{m,k} - 48k(m - k)^2(2m - k)\bar{\Delta}_{m,k-1}$$

$$-128k(m - k)(2m - k)(2m - k + 1)(m - k + 1)(k - 1)\bar{\Delta}_{m,k-2}, \tag{6.25}$$

starting with $\bar{\Delta}_{m,0} = 1$. Of course, the desired value $\mathcal{D}_m(1,0,0)$ is equal to $\bar{\Delta}_{m,m}$. An elementary computation with (6.25) modulo 6 shows that $\bar{\Delta}_{m,m} \equiv 3$ (mod 6) for all $m \in \mathbf{N}$. Hence $C_{m,0,0} \neq 0$.

We notice that $C_{0,m/2,0} = \mathcal{D}_m(0,1,0)$ if $m \equiv 0$ (mod 2) and $C_{0,0,m/3} = \mathcal{D}_m(0,0,1)$ if $m \equiv 0$ (mod 3). An easy recurrence shows

$$C_{0,m/2,0} = 864^m\big((m - 1)!!\big)^2(2m - 1)!!\,,$$

$$C_{0,0,m/3} = 864^m(-1)^{m/3}\binom{m}{m/3}\frac{(2m)!}{3^m}\,,$$

and hence these coefficients are also non-zero.

The coefficients $C_{1,0,\frac{m-1}{3}}$, defined for $m \equiv 1$ (mod 3) appear when we set $\bar{C} = 0$ in $\mathcal{D}_m(\bar{B},\bar{C},\bar{D})$ and skip the terms containing $\bar{B}^2$ and $\bar{B}^3$. Furthermore, we should include only one $\bar{B}$ factor. It is immediate to check that, except by the trivial factor $-18(4m^2 - 1)$, $C_{1,0,m-1/3}$ is the sum of all the determinants obtained from $\mathcal{D}_m(\bar{B},\bar{C},\bar{D})$ when we skip the row and column of index $3k+1$, for $k = 1,\ldots,\frac{m-1}{2}$, and we set $\bar{B} = \bar{C} = 0$, $\bar{D} = 1$. All the terms added have the sign of $(-1)^{\frac{m-1}{3}}$. Hence $C_{1,0,\frac{m-1}{3}} \neq 0$.

In a similar way, $C_{0,\frac{m-3}{2},1}$, defined for m odd, is found to be negative. Indeed, its value is the sum of all the determinants of the form $\mathcal{D}_m$ when we set $\bar{B} = 0$, $\bar{C} = 1$ and just one $\bar{D}$ of a row of odd index equal to 1, the others being zero.

To obtain $C_{0,1,\frac{m-2}{3}}$ we set in $\mathcal{D}_m$ the variable $\bar{B}$ equal to 0. Then one of the variables $\bar{C}$, in rows of index $3k + 2$, $k = 0,\ldots,\frac{m-2}{3}$, is set to 1 and the other $\bar{C}$'s are set to zero. $C_{0,1,\frac{m-2}{3}}$ is obtained by adding these determinants, and all of them have the sign of $(-1)^{\frac{m-2}{3}}$. Hence, it is non-zero.

Finally we proceed to show that $C_{0,2,\frac{m-4}{3}} \neq 0$, this coefficient being defined for $m \equiv 1 \pmod 3$. Set $\bar{B} = 0$ and consider all the possible choices of block structures for the matrix associated to $\mathcal{D}_m$, with the diagonals of the blocks contained in the diagonal of the initial matrix (that is: a block diagonal structure). We require that 2 blocks are 2×2 and the remaining ones are 3×3. In the 2×2 blocks set $\bar{C} = 1$ and in the 3×3 blocks set $\bar{C} = 0$, $\bar{D} = 1$. Then $C_{0,2,\frac{m-4}{3}}$ is the sum of all determinants $m \times m$ obtained in this way. The sign of all of them is the one of $(-1)^{\frac{m-4}{3}}$ and this ends the proof. $\square$

We have not been able to find an obvious proof that the remaining coefficients are different from zero. As for any specific problem they can be computed explicitly, we keep this as an assumption in the statement of the proposition.

For convenience we list here all the assumptions made on the $C_{i,j,k}$ coefficients and not proved before. We assume $m > 3$ and, of course, a coefficient $C_{i,j,k}$ is taken equal to zero if $j < 0$. Let

$$\beta(m) = (-1)^{\frac{m-4}{3}}$$
$$\times \left[C_{5,0,\frac{m-5}{3}} C^2_{0,1,\frac{m-2}{3}} - C_{3,1,\frac{m-5}{3}} C_{0,1,\frac{m-2}{3}} C_{2,0,\frac{m-2}{3}} + C_{1,2,\frac{m-5}{3}} C^2_{2,0,\frac{m-2}{3}} \right],$$

be defined for $m \equiv 2 \pmod 3$, and

$$\gamma(m) = C_{0,\frac{m-3}{2},1} \sum_{k=0}^{5} C_{5-k,\frac{m-5}{2}-k,k}\, \mu^k$$
$$- \left(\sum_{k=0}^{3} C_{3-k,\frac{m-3}{2}-k,k}\, \mu^k \right) \left(\sum_{k=1}^{3} k\, C_{3-k,\frac{m-3}{2}-k,k}\, \mu^{k-1} \right),$$

where $\mu = -C_{1,\frac{m-1}{2},0}/C_{0,\frac{m-3}{2},1}$, defined for m odd.

Then the assumptions of Proposition 6.2 are: $\beta(m)$ and $\gamma(m)$ must be non-zero whenever defined. These assumptions have been tested up to $m = 1000$ and they are satisfied in all cases. The check has been done by constructing a specific program for symbolic computation. Furthermore, beyond an eventual

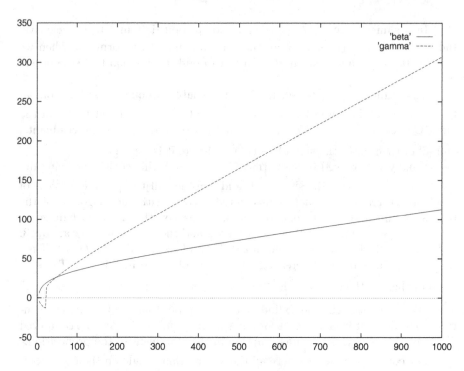

Figure 6.1: The coefficients β and γ as functions of m for m between 4 and 1000. For scaling reasons we have replaced B, C, D by $2B, C/2, D/8$, respectively, and the coefficients $C_{i,j,k}$, entering in the expression of β and γ have been divided by $(36(1 - 4m^2))^m$, where $m = i + 2j + 3k$ is the weight of the coefficient, before the computation of β and γ. For each one of these two functions one plots the respective arg sh of β and γ as a function of m.

transient for small values of m the functions $\beta(m)$ and $\gamma(m)$ seem to increase very quickly in a quite regular way. Figure 6.1 displays this behaviour. Then we state the following conjecture.

Conjecture *For all $m > 3$, when the functions β and γ are defined, they are non-zero.*

We summarize all the results of this section in the following theorem.

Theorem 6.2 *Assume that a classical Hamiltonian system with a potential like (6.3) has NVE of Lamé type associated to the family of solutions, Γ_h, lying on the $x_2 = 0$ plane and parametrized by the energy, h. Then, a necessary* *condition for integrability is that the related polynomials P_1 and P_2 (see Section 6.1) satisfy $a_2 = 0$, and one of the following conditions holds:*

1. $a_1 = \frac{4}{n(n+1)}$ *for some* $n \in \mathbf{N}$,

2. $a_1 = \frac{16}{4m^2-1}$ *for some* $m \in \mathbf{N}$. *Then, assuming the conjecture above is true, one should have* $b_2 = 0$ *and we should be in one of the following cases:*

   2.1) $m = 1$ *and* $b_1 = 0$,

   2.2) $m = 2$ *and* $c_2 = 0$, $16a_1c_1 + 3b_1^2 = 0$,

   2.3) $m = 3$ *and* $16a_1d_2 + 11b_1c_2 = 0$, $1024a_1^2d_1 + 704a_1b_1c_1 + 45b_1^3 = 0$,

   2.m) $m > 3$. *Then we should have* $b_1 = 0$ *and, furthermore, either* $c_1 = c_2 = 0$ *if* m *is congruent with* $1, 2, 4$ *or* 5 *modulo* 6, *or* $d_1 = d_2 = 0$ *if* m *is odd.*

3. $a_1 = \frac{4}{n(n+1)}$ *with* $n + \frac{1}{2} \in \frac{1}{3}\mathbf{Z} \cup \frac{1}{4}\mathbf{Z} \cup \frac{1}{5}\mathbf{Z}\backslash\mathbf{Z}$, $b_2 = 0$ *and either* $c_2 = 0, b_1^2 - 3a_1c_1 = 0$ *or* $c_2b_1 - 3a_1d_2 = 0$, $2b_1^3 - 9a_1b_1c_1 + 27a_1^2d_1 = 0$.

We recall that, for a Lamé type equation, it is necessary to have discriminant $\Delta(h) \neq 0$. If $\Delta(h) \equiv 0$ we have several conditions on the coefficients of P_1 and P_2. If we denote by $\hat{A} = a_1 + h\,a_2, \ldots, \hat{D} = d_1 + h\,d_2$, then either $\hat{A}(h) \equiv 0$ or $R(h) \equiv 0$, where $R(h) = 27\,\hat{A}^2\hat{B}^2 - 18\,\hat{A}\,\hat{B}\,\hat{C}\,\hat{D} + 4\,\hat{A}\,\hat{C}^3 + 4\,\hat{B}^3\hat{D} - \hat{B}^2\hat{C}^2$.

6.3 Examples

We shall consider different examples belonging to families A), B) and C) of Section 6.1.

(1) *Cubic potentials.* They appear as family A). In that case $d_2 > 0$, $\alpha = \pm\sqrt{\frac{d_2}{2}}x_1 + e$ and $P_1(\alpha) = a_1\alpha^3 + b_1\alpha^3 + c_1\alpha + d_1$, with $a_1 \neq 0$, gives immediately the potential by (6.7). We remark that *all* the cubic potentials of the form (6.3) having *NVE* of Lamé type associated to $x_2 = 0$ appear in this way. From Remark 3 it follows that the discriminant condition is always satisfied.

From Theorem 6.2 necessary conditions for integrability are that some of the following holds:

1. $a_1 = \frac{4}{n(n+1)}$, $n \in \mathbf{N}$,

2. $a_1 = \frac{16}{4m^2-1}$ and then: if $m = 1$, $b_1 = 0$; if $m = 2$, $16a_1c_1 + 3b_1^2 = 0$; if $m > 3$ and m is congruent to $1, 2, 4$ or 5 (mod 6), $b_1 = c_1 = 0$. Other values of m give non-integrability,

3. $a_1 = \frac{4}{n(n+1)}$, with $n + \frac{1}{2} \in \frac{1}{3}\mathbf{Z} \cup \frac{1}{4}\mathbf{Z} \cup \frac{1}{5}\mathbf{Z}\backslash\mathbf{Z}$ and then $b_1^2 = 3c_1a_1$.

As a concrete example we can apply this to the generalized Hénon-Heiles potentials studied by Ito using Ziglin's theorem [48]. This family of potentials is given by

$$V(x_1, x_2) = \frac{1}{2}x_1^2 + \frac{1}{3}dx_1^3 + (\frac{1}{2} + cx_1)x_2^2, \qquad (6.26)$$

with $c \neq 0$, d parameters (for $c = 0$ the above potential is separable).

As Ito showed, by a rotation in the configuration space

$$x_1 = 1/\sqrt{3 - d/c}\,(\hat{x}_1 + \sqrt{2 - d/c}\,\hat{x}_2),$$
$$x_2 = 1/\sqrt{3 - d/c}\,(\hat{x}_2 - \sqrt{2 - d/c}\,\hat{x}_1),$$

we get a canonical transformation, provided $d/c \neq 3$. The transformed potential is given by

$$\hat{V}(\hat{x}_1, \hat{x}_2) = \frac{1}{2}\hat{x}_1^2 + 2c/\sqrt{3 - d/c}\,\hat{x}_1^3 + (\frac{1}{2} + (d - c)/\sqrt{3 - d/c}\,\hat{x}_1)\hat{x}_2^2 + O(\hat{x}_2^3), \qquad (6.27)$$

$\hat{y}_1$ and $\hat{y}_2$ being the conjugated moments.

We can apply our results to both potentials, and the integrability of the potential (6.26) is only compatible with the integrability of V and $\hat{V}$ at the same time (in other words, the potential (6.26) has two invariant planes $x_2 = y_2 = 0$ and $\hat{x}_2 = \hat{y}_2 = 0$ as in some examples of Chapter 5).

A simple computation recovers the Ito result ([48], Theorem 3).

Proposition 6.3 *Except for $c/d = 0, 1, 1/6, 1/2$ the Hamiltonian system defined by the potential (6.26) is non-integrable with meromorphic first integral.*

For $c/d = 0, 1, 1/6$ the system is integrable.

For $c/d = 1/2$ numerical simulations suggest the non-integrability of the system, but a rigorous proof is still missing. We remark that the value $c/d = 1/2$ is in some sense special. For instance, for this value the two above invariant planes coincide. We will come back to this system in Chapter 8.

(2) *Quartic potentials.* Assume we are in the family B) case with $P_2(\alpha) = c_2\alpha + d_2$, $c_2 \neq 0$. We recall that then $\alpha(x_1) = \frac{c_2}{8}x_1^2 + e\,x_1 + \frac{2e^2 - d_2}{c_2}$, e being arbitrary. Assume $P_1(\alpha) = P_2(\alpha)\,S(\alpha)$, where S is a polynomial of degree two. Then the potential V is quartic (if in the $O(x_2^3)$ terms we only include $x_2^3, x_1x_2^2, x_2^4$). Furthermore this is the only way in which quartic potentials of the form (6.3) can be obtained. Let $S(\alpha) = s_2\alpha^2 + s_1\alpha + s_0$ and $\alpha(x_1) = \alpha_2x_1^2 + \alpha_1x_1 + \alpha_0$, by relabelling the coefficients. We note that $s_2, s_1, s_0, \alpha_2, \alpha_1, \alpha_0$ are arbitrary provided $s_2 \neq 0$, $\alpha_2 \neq 0$. This, together with the arbitrariness of the coefficients of $x_2^3, x_1x_2^3$ and x_2^4, is all the freedom available to have a quartic potential of the form (6.3) with *NVE* of Lamé type. Notice that *not* all the

quartic potentials of the form (6.3) appear in this way. Only a codimension-two subfamily. The coefficients of $P_1(\alpha)$ are $a_1 = c_2 s_2$, $b_1 = c_2 s_1 + d_2 s_2$, $c_1 = c_2 s_0 + d_2 s_1$, $d_1 = d_2 s_0$. From Remark 3 it follows that the discriminant condition is always satisfied.

From Theorem 6.2, the necessary conditions for integrability are that one of the following holds:

1. $c_2 s_2 = \frac{4}{n(n+1)}$, $n \in \mathbf{N}$,

2. $c_2 s_2 = \frac{16}{3}$ and $c_2 s_1 + d_2 s_2 = 0$, or $c_2 s_2 = \frac{16}{4m^2-1}$, $m \geq 3$, m odd, and $d_2 = s_1 = 0$.

3. $c_2 s_2 = \frac{4}{n(n+1)}$, $n + \frac{1}{2} \in \frac{1}{3}\mathbf{Z} \cup \frac{1}{4}\mathbf{Z} \cup \frac{1}{5}\mathbf{Z}\backslash\mathbf{Z}$ and $c_2 s_1 = 2d_2 s_2$.

(3) *Rational potentials.* Again in the case of family B) and with P_2 and α as before, assume that $P_1(\alpha) = a_1\alpha^3 + b_1\alpha^2 + c_1\alpha + d_1$ cannot be divided by $P_2(\alpha)$. Then $\varphi(x_1)$ is a rational function, quotient of a polynomial of degree 6 by one of degree 2. The subfamily of rational functions of this type that can be obtained has codimension 3. Furthermore, when the rational function is given, the terms containing x_2^2 are fixed (except by multiplicative constants).

Integrability conditions are immediate from Theorem 6.2. We only remark that case 2.2) can not occur and case 2.m) can only occur with m odd.

(4) *Periodic Toda lattice with 3 particles and two equal masses.* Given the Hamiltonian with 3 degrees of freedom

$$H_1 = \frac{1}{2}\left(\frac{p_1^2}{m} + p_2^2 + p_3^2\right) + e^{q_1 - q_2} + e^{q_2 - q_3} + e^{q_3 - q_1},$$

which by means of the center of mass reduction can be simplified to

$$H = \frac{1}{2}(y_1^2 + y_2^2) + 2\,e^{2x_1}\cosh\left(2\sqrt{3\mu}\,x_2\right) + e^{-4x_1},$$

where μ is defined by $m = \frac{2}{3\mu-1}$. This potential is a particular case of C1) with

$$\alpha = -24\mu\,e^{2x_1},\ e = -24\mu,\ c_2 = d_2 = 0,$$

and

$$P_2(\alpha) = 8\alpha^2,\ P_1(\alpha) = \frac{2}{3\mu}\alpha^3 - 8(24\mu)^2.$$

Hence $n(n+1) = \frac{4}{a_1} = 6\mu$. If $n \notin \mathbf{N}$ the system is non-integrable because $b_2 \neq 0$.

(5) *Potential on a cylinder coinciding locally with Hénon-Heiles.* Consider $x_1 \in \left(-\frac{\pi}{2}, \frac{\pi}{2}\right)$, $x_2 \in S^1$ and the potential

$$V = \frac{1}{2}D\sin x_1\,\mathrm{tg}^2 x_1 + \frac{1}{2}\mathrm{tg}^2 x_1 + \frac{1}{2}\left(C\sin x_1 + 1\right)\sin^2 x_2.$$

It coincides with the Hénon-Heiles potential around (0,0) up to third order. It is of the type C.3) with $a_1 = \frac{2D}{3C}$, $b_2 = -2$. Hence, if $\frac{C}{D} \neq \frac{n(n+1)}{6}$, $n \in \mathbf{N}$, it is non-integrable.

6.4   The homogeneous Hénon-Heiles potential

We recall that, from Chapter 5, the Hamiltonian of the homogeneous Hénon-Heiles potential is given by

$$H = \frac{1}{2}\left(y_1^2 + y_2^2\right) + \frac{e}{3}x_1^3 + x_1x_2^2. \tag{6.28}$$

We can derive the integrability conditions from the general analysis of cubic potentials in Section 6.3, but we shall proceed directly. We recall some facts of this Hamiltonian. By a suitable rotation in the configuration space we get the Hamiltonian

$$H = \frac{1}{2}(\eta_1^2 + \eta_2^2) + \frac{2}{3(e-1)}\xi_1^3 + \xi_1\,\xi_2^2 + \frac{\sqrt{2-e}}{3}\,\frac{e+1}{e-1}\,\xi_2^3, \tag{6.29}$$

if $e \neq 1$ (if $e = 1$, it reduces to a separable potential). Then as we have already remarked in Chapter 5, (6.29) is as (6.28) with $\hat{e} = \frac{2}{e-1}$ instead of e.

The first goal of this section is to give a new proof of the Proposition 5.4.

Proposition 6.4 *The Hamiltonian* (6.28) *is non-integrable for* $e \in \mathbf{C}\backslash\{1,2,6,16\}$.

Proof. Leaving aside the case $e = 2$, we derive the following conditions for the coefficient n: $n(n+1) = \frac{12}{e}$. If we denote by $\hat{n}$ the coefficient associated to $\hat{e} : \hat{n}(\hat{n}+1) = \frac{12}{\hat{e}}$, from the relation between e and $\hat{e}$ we have

$$6\left(\frac{12}{n(n+1)} - 1\right) = \hat{n}(\hat{n}+1). \tag{6.30}$$

To be in one of the cases i), ii) or iii) of Subsection 2.8.4, both n and $\hat{n}$ must be rationals with denominator 1, 2, 4, 6 or 10. If $n \neq 0$ (the case $n = 0$ corresponding to $e = \infty$) from (6.30) we have

$$\left|\hat{n}(\hat{n}+1)\right| \leq 6\left(\frac{12}{\frac{1}{10}\cdot\frac{9}{10}} + 1\right) = 806.$$

Therefore $|\hat{n}|$ has an upper bound and it remains to examine a finite number of cases. A direct check shows that the only possible solutions of (6.30), with the required conditions, are:

1. $n = 3$, $\hat{n} = 0$, corresponding to $e = 1$,
2. $n = 2$, $\hat{n} = 2$, corresponding to $e = 2$ (notice that in this case $\hat{e}$ is also equal to 2),
3. $n = 1$, $\hat{n} = 5$, corresponding to $e = 6$,
4. $n = \frac{1}{2}$, $\hat{n} = 9$, corresponding to $e = 16$.

This ends the proof. $\square$

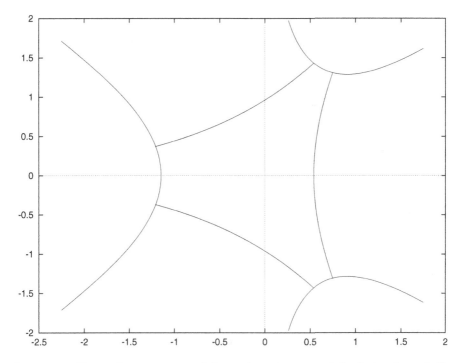

Figure 6.2: Zero velocity curve and (x_1, x_2) projections of the 3 simple periodic orbits, for $e = -2$.

Now we discuss the dynamical meaning of non-integrability. For $e < 0$ it is easy to show that there are 3 simple periodic orbits, all of them touching the zero velocity curve (zvc) in two points. One of them is symmetrical: one can take $x_2 = y_1 = 0$ as initial conditions. Due to the homogeneity it is enough to consider energy level $h = 1$. Figure 6.2 displays the zvc for $e = -2$ and also the 3 simple periodic orbits (γ_1, symmetrical and γ_2, γ_3, symmetric to each other) projected on the (x_1, x_2) plane. These orbits are hyperbolic. For $e \nearrow 0$ the eigenvalue of largest modulus of γ_1 tends to 1 and those of γ_2, γ_3 to ∞. For $e \searrow -\infty$ the eigenvalue of γ_1 tends to ∞ and those of γ_2, γ_3 to 1.

Figure 6.3 shows the intersection of γ_1 and its unstable and stable manifolds with the Poincaré section $x_2 = 0$. The boundary of the Poincaré section is for $y_2 = 0$ and, hence it is given by the cubic $\frac{y_1^2}{2} + \frac{e}{3}x_1^3 = 1$ (see Figure 6.3). The invariant manifolds intersect transversally at an homoclinic point and this implies chaotic dynamics. Similar patterns appear for any $e < 0$ (but they are difficult to see for $|e|$ small, for instance).

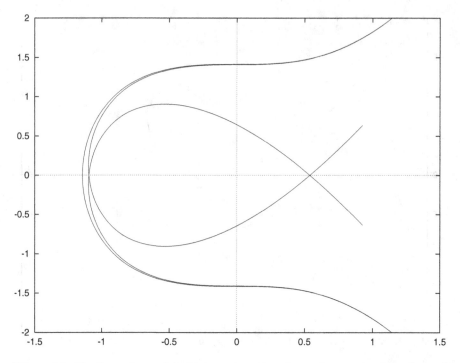

Figure 6.3: The boundary of the Poincaré section through $x_2 = 0$ on the (x_1, y_1) coordinates and the sections of the invariant manifolds of γ_1 for $e = -2$ and $h = 1$.

The case $e \geq 0$ is more subtle. One has $\ddot{x}_1 = -ex_1^2 - x_2^2$ and, as the x_1 acceleration is always $\ddot{x}_1 \leq 0$, there is no possible recurrence in the real phase space. We can look for it in the complex phase space. Let 1_α denote a complex number of modulus 1 and argument α.

The changes

$$x_1 = 1_{\pi/3}\xi_1 \,, \ x_2 = 1_{-\pi/6}\,\xi_2 \,, \ y_1 = 1_{-\pi/2}\,\eta_1 \,, \ y_2 = \eta_2 \,, \ t = 1_{-\pi/6}\,s$$

lead to the Hamiltonian (using s as new time)

$$H = \frac{1}{2}\big(-\eta_1^2 + \eta_2^2\big) + \frac{e}{3}\big(-\xi_1^3\big) + \xi_1\xi_2^2 \,, \tag{6.31}$$

which is real for real variables, and on the same level of energy. For this case the acceleration changes sign. Take, for instance, the value $e = 3/2$. Figure 6.4 shows a Poincaré section of (6.31) on $h = 1$, $\xi_2 = 0$. One can see a typical

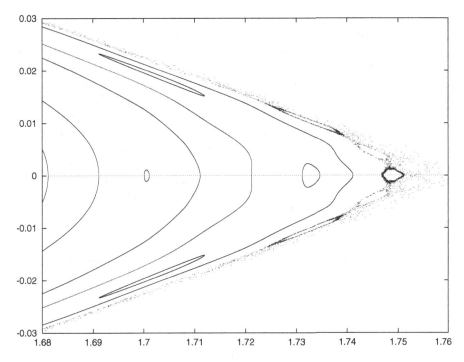

Figure 6.4: A part of the phase portrait of the Poincaré map through $\xi_2 = 0$ on the (ξ_1, η_1) coordinates for $e = 1.5$ and $h = 1$.

pattern with invariant curves, islands and chaotic regions. Symmetric periodic orbits appear for $\xi_1^e = 1.346146\ldots$ (elliptical) and $\xi_1^h = 1.766010\ldots$ (hyperbolic). In Figure 6.4 initial points are taken with $\xi_1 = \xi_1^e + \delta$, $\eta_1 = 0$, where $\delta = 0.335(0.01)0.405$. In fact, the last of these points is close to the hyperbolic one and escapes after a few thousands of iterates of the Poincaré map. A similar behaviour is observed for other values of e. However, for e near to 1.51602386 the elliptic and hyperbolic simple symmetric periodic orbits coincide in a parabolic orbit and they can not be continued to larger values with real (ξ, η) variables (of course, there are no periodic orbits with real (x, y) variables). We are interested in the case $e = 2$.

It is not too difficult to move e to the complex plane, to follow a path and obtain the corresponding symmetric ($x_2 = 0$ and $y_1 = 0$ initially) complex periodic orbits (of real dimension 1), with complex period. We remark that the passage from e to the initial value x_1^0 is an interesting Riemann surface (we keep $h = 1$), having branches when the eigenvalues are equal to 1.

For $e = 2$ the initial value $x_1^0 \simeq 0.01247621 + i\,1.08807831$, leads to a periodic orbit with period $T \simeq 5.25449302 - i\,2.60364041$ and dominant eigenvalue $\lambda \simeq -3.39790418 + i\,27.26367123$. We keep the Poincaré section $x_2 = 0$ (as complex). The unstable and stable invariant manifolds have been generated, numerically, from a fundamental domain (in the Poincaré section) diffeomorphic to a real 2D annulus. The intersections of these manifolds with this section on $h = 1$ are complex symmetric curves (of real dimension 2). If $x_1 = g_u(y_1)$, $x_1 = g_s(y_1)$ describe, locally, the unstable and stable manifolds respectively, one has $g_s = -g_u$. A homoclinic point has been found near $x_1 = 1.15441741 + i\,0.14795498$, $y_1 = 0$. At this point one has $\frac{dx_1}{dy_1} \simeq -0.0514 + 0.1042\,i$. Hence the manifolds intersect transversally and we get chaotic dynamics. In particular this prevents the integrability for $e = 2$.

We note the strong analogy with the Hénon-Heiles generalized potential studied in Section 6.3. The parameter e here corresponds to the parameter d/c there.

The preceeding discussion leads to the following natural questions. The first one is about the homogeneous Hénon-Heiles potential.

Question 1. *Is the Hénon-Heiles homogeneous potential integrable for all $e > 0$ if we restrict to the real phase space?*

The second question is of a more general character and very much related to some comments that we made in Chapters 3 and 4.

Question 2. *Is it true that if a system is non-integrable, (in the sense used in this monograph), chaotic dynamics occurs in some part (eventually complex) of the phase space?*

We shall come back to Question 2 in Chapters 7 and 8.

Chapter 7

A Connection with Chaotic Dynamics

The motivation of this chapter is to clarify the relation between the *real* chaotic dynamics of non-integrable Hamiltonian systems and the purely algebraic differential Galois criterion of non-integrability based on the analysis in the *complex* phase space of the variational equations along a particular integral curve. This problem was proposed in Section 6.4 (Question 2).

Concretely, and as a first step to understanding the above problem, we consider the relatively simple situation of a two degrees of freedom Hamiltonian system with a (real) homoclinic orbit contained in an invariant plane and asymptotic to a center-saddle equilibrium point. In this situation Lerman [64] gives a necessary criterion, in terms of some kind of asymptotic monodromy matrix of the normal variational equations along the homoclinic orbit, for the non-existence of transversal homoclinic orbits associated to the invariant manifolds of the Lyapounov orbits around the equilibrium point, i.e., real "dynamical integrability" in a neighborhood of the homoclinic orbit (Theorem 3.5). This condition was interpreted by Grotta-Ragazzo [40] in terms of a global monodromy matrix of the algebraic normal variational equation *ANVE* in the complex phase space and, as the author (see the Introduction), he also conjectured the existence of some connection between the Lerman's theorem and Ziglin's non-integrability criterion (Theorem 3.4). The present chapter is devoted to clarification of this relation. Instead of Ziglin's original theorem we prefer to work with the more general theory in terms of the differential Galois group of the variational equations. We recall that we proved in Corollary 4.6 that, for two degrees of freedom Hamiltonian systems, Ziglin's Theorem is a corollary in our theory.

The main result of this chapter is that (under suitable assumptions of complex analycity) the two above necessary conditions for integrability are indeed the same, when we restrict the analysis to a complex neighborhood of

J. J. Morales Ruiz, *Differential Galois Theory and Non-Integrability*   131
of Hamiltonian, Systems, Modern Birkhäuser Classics,
DOI: 10.1007/978-3-0348-0723-4_7, © Springer Basel 1999

the real homoclinic orbit. This explains why in some applications there exist two different proofs of non-integrability: one based on the Lerman theorem and the other on the Ziglin theorem. The important thing is that, in this case, an "algebraic" obstruction to integrability is also a chaotic obstruction to integrability. Section 7.3 is devoted to a detailed analysis of an example with a normal variational equation of Lamé type.

The results of this chapter were obtained in a joint work of the author with J.M. Peris [76].

7.1   Grotta-Ragazzo interpretation of Lerman's theorem

In his interpretation of Lerman's theorem, Grotta-Ragazzo considered the (complexified) *NVE* along the (complex) homoclinic orbit Γ contained in the invariant plane $x_2 = y_2 = 0$ (Γ: $x_1 = x_1(t)$, $y_1 = y_1(t)$, $x_2 = y_2 = 0$)

$$\ddot{\xi} + \alpha(x_1(t))\xi = 0,$$

where the Hamiltonian is given by (3.1).

By the change of independent variables $x := x_1(t)$, he obtain the *ANVE*,

$$\frac{d^2\xi}{dx^2} + \frac{\varphi'(x)}{2\varphi(x)}\frac{d\xi}{dx} - \frac{\alpha(x)}{2\varphi(x)}\xi = 0. \tag{7.1}$$

Among the real singularities of this variational equation are the equilibrium point $x_1 = x = 0$ and the branching point (of the covering $t \to x$)

$$x_1 = x = e, y_1 = (2\varphi(e))^{1/2} = 0,$$

corresponding to the zero velocity point of the homoclinic orbit. Let σ be a closed simple arc (element of the fundamental group) in the (complex) x-plane surrounding only the singularities $x = 0$, $x = e$, and m_σ the monodromy matrix of the above equation along σ. Then Grotta-Ragazzo obtained the following result ([40], Theorem 8).

Theorem 7.1 *The matrix R in Lerman's theorem (Theorem 3.5) is a rotation if, and only if, $m_\sigma^2 = 1$ (identity).*

In his proof Grotta-Ragazzo used the relation between the monodromy m_σ and the reflexion coefficient of the *NVE* (considered as a Schrödinger equation).

7.2 Differential Galois approach

In this section we shall give the relation between the Grotta-Ragazzo result presented in the last section and the differential Galois obstruction to integrability of Chapter 4. In order to get this, we start with a reformulation of Theorem 7.1.

So, let X_H be a two degrees of freedom Hamiltonian system with a saddle-center equilibrium point and a homoclinic orbit $\Gamma_{\mathbf{R}}$ to this point contained in an invariant plane $x_2 = y_2 = 0$.

We consider this real Hamiltonian system as the restriction to the real domain of a complex analytic Hamiltonian system (with complex time), as in some examples in the chapters above. If we add the origin to the homoclinic orbit, then we get a complex analytic singular curve. Now we can particularize the general constructions of Chapter 4 to this. The origin is the singular point and by desingularization one obtains a non-singular (in a neighborhood of the origin) analytic curve $\overline{\Gamma}$. On $\overline{\Gamma}$ there are two points, 0^+ and 0^-, corresponding to the origin. We note that the homoclinic orbit is, up to first order, defined by $x_1 y_1 = 0$, while the desingularized curve is defined by the pair of disconnected lines $x_1 = 0$, $y_1 = 0$ with two points at the origin. These points are, in the temporal parametrization, $t = +\infty$ and $t = -\infty$.

We are interested only in a domain $\overline{\Gamma}_{\mathrm{loc}}$ of the Riemann surface $\overline{\Gamma}$ that contains $\Gamma_{\mathbf{R}}$ and the points 0^+ and 0^-. This Riemann surface $\overline{\Gamma}_{\mathrm{loc}}$ is parametrized by three coordinate charts A_-, A_t and A_+ with coordinates $x := x_1$, t and $y := y_1$ respectively. Then, by restriction to a small enough domain, it is always possible to get a Riemann surface $\overline{\Gamma}_{\mathrm{loc}}$ on which the only singularities of the *NVE* are 0^+ and 0^-.

Let γ be the closed simple path in $\overline{\Gamma}_{\mathrm{loc}}$ surrounding $\Gamma_{\mathbf{R}}$. If we denote by m_γ the corresponding monodromy matrix of the *NVE*, then by the double covering $t \to x$ of the above section, we have $m_\gamma = m_\sigma^2$. Hence, by the Grotta-Ragazzo theorem we get the following.

Proposition 7.1 *The matrix R in Theorem 3.5 is a rotation if, and only if,* $m_\gamma = 1$.

In order to obtain the connection with the Galois Theory, we need to do some elementary analysis on the algebraic groups of $SL(2, \mathbf{C})$ generated by hyperbolic elements.

Lemma 7.1 *Let M be a subgroup of $SL(2, \mathbf{C})$ generated by k elements m_1, m_2, ..., m_k, such that each m_i has eigenvalues $(\lambda_i, \lambda_i^{-1})$ with $|\lambda_i| \neq 1$, $i = 1, 2, \ldots, k$. Then the closed group $\overline{M}$ (in the Zariski topology) must be one of the groups 4, 6 or 7 of Proposition 2.2.*

Proof. As $\overline{M}$ is an algebraic subgroup of $SL(2, \mathbf{C})$, it is one of the groups 1–7 in Proposition 2.2. We analyze each of these cases. The group M is not a finite group, as m_i has infinite order. Also m_i is not contained in the triangular groups of types 2 or 3 of Proposition 2.2, because the eigenvalues of all the elements of these groups have eigenvalues on the unit circle.

Finally, if we are in case 5, necessarily $m_i \in G^0$ (because the eigenvalues of $G \backslash G^0$ are in the unit circle). But then $M \subset G^0$ (G^0 is a group) and $\overline{M} \neq G$ ($\overline{M}$ is the smallest algebraic group that contains M, and furthermore G^0 is an algebraic group). $\square$

As we shall show, this elementary result is central to our analysis.

Now we come back to the local homoclinic complex orbit Γ_{loc} with the two singularities 0^+, 0^-, coming from the equilibrium point, with monodromy matrices m_+, m_-. Let $m_\gamma = m_+ m_-$ be the monodromy around the two singular points. Let G_{loc} be the Galois group of the *NVE* restricted to Γ_{loc} (this is a linear differential equation with meromorphic coefficients over the simply connected domain of the complex plane $\overline{\Gamma}_{\mathrm{loc}}$ obtained by adding to Γ_{loc} the two singular points 0^+ and 0^-, i.e., we consider the differential field of coefficients of the *NVE* as the meromorphic functions over $\overline{\Gamma}_{\mathrm{loc}}$: see Section 2.2). Notice that, for simplicity of notation, here we have used the same notation for this "local" Galois group and the local Galois group at a singular point (Chapter 2), but they are different objects. Although the philosophy in both cases is the same: we consider a bigger coefficient field and, by Theorem 2.2, we get a smaller Galois group.

As a direct consequence of the lemma above we obtain the following proposition.

Proposition 7.2 *The monodromy matrix m_γ is equal to the identity if, and only if, the identity component $(G_{\mathrm{loc}})^0$ is abelian. Furthermore, in this case, the Galois group is of type 4 in Proposition 2.2.*

Proof. If $m_+ m_- = \mathbf{1}$ it is clear that the monodromy group M is abelian, for the monodromy group is generated by a single element (for instance, m_+). As the equation is of Fuchs type, then $\overline{M} = G_{\mathrm{loc}}$ is abelian and of type 4 in Proposition 2.2 (as the reader can verify, the Zariski closure of the group generated by a diagonal matrix of infinite order in $SL(2, \mathbf{C})$ is always of type 4).

Reciprocally, we know that the monodromy group has two generators m_+, m_- with inverse eigenvalues lying outside of the unit circle. By Lemma 7.1, if the identity component $(G_{\mathrm{loc}})^0$ is abelian, then as $\overline{M} = G_{\mathrm{loc}}$, $_{\mathrm{loc}}G$ (and in fact the global Galois group G) is of type 4. Furthermore, from the fact that the two matrices m_+, m_- have inverse eigenvalues ($\lambda_+ = \lambda_-^{-1}$), ($\lambda_+, \lambda_+^{-1}$) and ($\lambda_-, \lambda_-^{-1}$) being the eigenvalues of m_+ and m_- respectively, we get the desired result. $\square$

In this way, we have proved that two unrelated first order obstructions to integrability are, in fact, the same (under suitable assumptions of analyticity). The first one, given by the condition in Lerman's theorem, has been formulated in terms of real dynamics (the existence of transversal homoclinic orbits). The second one is formulated in an algebraic way (Corollary 4.5 restricted to $\overline{\Gamma}_{\mathrm{loc}}$) and has a meaning in the complex setting only. Summarizing, we have obtained the following differential Galois interpretation of the Lerman and Grotta-Ragazzo results.

Theorem 7.2 *If the identity component* $(G_{\mathrm{loc}})^0$ *is not abelian, then there exists no additional meromorphic first integral in a neighborhood of* Γ_{loc} *and the invariant manifolds of the Lyapounov orbits must intersect transversally.*

7.3 Example

We shall apply Theorem 7.2 to a two degrees of freedom potential with a *NVE* of Lamé type.

We take, in the formulas of Chapter 3, (3.1) and (3.2),

$$\varphi(x_1) = -\frac{1}{2e_1e_2}x_1^2(x_1 - e_1)(x_1 - e_2), \tag{7.2}$$

$$\alpha(x_1) = \frac{3}{16e_1e_2}x_1^2 + ax_1 + b, \tag{7.3}$$

where we normalize $\nu = 1$ and $b := \omega^2$. So, the system depends on four (real) parameters e_1, e_2, a, b, with $e_1 \neq e_2$ and $b > 0$.

We are going to compute for this example $\Gamma_{\mathbf{R}}$, C, Γ and $\overline{\Gamma}$ (from these constructions we get Γ_{loc} and $\overline{\Gamma}_{\mathrm{loc}}$).

The (complex) analytical curve C is $y_1^2 + 2\varphi(x_1) = 0$ $(x_2 = y_2 = 0)$. Without loss of generality we assume $e_1 > 0$ and then either $0 < e_1 < e_2$ or $e_2 < 0 < e_1$. In both cases we can take $\Gamma_{\mathbf{R}}$ as the unique real homoclinic orbit contained in C, $0 < x_1 \leq e_1$ (the canonical change $x_1 \to -x_1$, $y_1 \to -y_1$, reduces all the possibilities to the above one). The complex orbit Γ is C minus the origin (we recall that the temporal parameter t is a local parameter on Γ).

The desingularized curve $\overline{\Gamma}$ is the projective line $\mathbf{P}^1$. In fact, by the standard birational change $x_1 = \tilde{x}$, $y_1 = \tilde{x}\tilde{y}/\sqrt{e_1e_2}$, we get the genus zero curve $\tilde{y}^2 = (\tilde{x}-e_1)(\tilde{x}-e_2)$. Now, with the change $\tilde{y} = \frac{1}{2}(e_1-e_2)\hat{y}$, $\tilde{x} = \frac{1}{2}(e_1-e_2)\hat{x}+\frac{e_1+e_2}{2}$, we obtain the curve $\hat{x}^2 - \hat{y}^2 = 1$. This last curve is parametrized by the rational functions

$$\hat{x} = \frac{r^2+1}{r^2-1}, \qquad \hat{y} = 2\frac{r}{r^2-1}.$$

If we compose all these changes we obtain a rational r-parametrization of $\overline{\Gamma}$. So, $\overline{\Gamma} = \Gamma \cup \{r = \pm\sqrt{e_2/e_1},\ r = \pm 1\}$, $r = \pm\sqrt{e_2/e_1}$ being the two points corresponding to the origin $x_1 = y_1 = 0$, and $r = \pm 1$ the two points at infinity. It is interesting to express $\Gamma_{\mathbf{R}}$ in this parametrization:

$$\Gamma_{\mathbf{R}} = \left\{ r \in \mathbf{C} : \mathrm{Re}(r) = 0, |r| > \sqrt{-\frac{e_2}{e_1}} \right\} \cup \{\infty\} \quad \text{if} \quad e_1 e_2 < 0,$$

$$\Gamma_{\mathbf{R}} = \left\{ r \in \mathbf{C} : \mathrm{Im}(r) = 0, |r| > \sqrt{\frac{e_2}{e_1}} \right\} \cup \{\infty\} \quad \text{if} \quad e_1 e_2 > 0.$$

Then, the *NVE* in these coordinates is

$$\frac{d^2\eta}{dr^2} + P\frac{d\eta}{dr} + Q\eta = 0, \tag{7.4}$$

where $P = 2\frac{re_1}{e_1 r^2 - e_2}$ and $Q = (Cr^4 - Dr^2 + E)/(4(e_1 r^2 - e_2)^2(r^2 - 1)^2)$, with $C = 16ae_1^2 e_2 + 16be_1 e_2 + 3e_1^2$, $D = 16ae_1^2 e_2 + 32be_1 e_2 + 16ae_1 e_2^2$, $E = 16be_1 e_2 + 3e_2^2 + 16ae_1 e_2^2$.

We know from the general theory that this equation is symplectic. In other words, its Galois Group is contained in $SL(2, \mathbf{C})$. Indeed, it is easy to check this in a direct way; so, $P = \frac{d}{dr}\log(e_1 r^2 - e_2)$ (see Section 2.2). Furthermore, their singularities are $r = \pm 1$ (with difference of exponents $1/2$) and $r = \pm\sqrt{e_2/e_1}$ (with exponents $\pm i\sqrt{b}$). From this, and from the symmetry in r it follows that it is possible to reduce this equation to a Lamé differential equation, if we take r^2 as the new independent variable. But we prefer to make this reduction in a more standard way.

So, by the covering $\overline{\Gamma} = \mathbf{P}^1 \to \mathbf{P}^1$ ($r \mapsto x = x_1$), we obtain the *ANVE* (7.1),

$$\frac{d^2\xi}{dx^2} + \left(\frac{1}{x} + \frac{\frac{1}{2}}{x - e_1} + \frac{\frac{1}{2}}{x - e_2} \right)\frac{d\xi}{dx} + \frac{\frac{3}{4}x^2 + 4e_1 e_2 ax + 4e_1 e_2 b}{4x^2(x - e_1)(x - e_2)}\xi = 0. \tag{7.5}$$

In order to show that this equation is of Lamé type, it is necessary to make some transformations. First, if we take $z = 1/x$, we get

$$\frac{d^2\xi}{dz^2} + \left(\frac{\frac{1}{2}}{z - s_1} + \frac{\frac{1}{2}}{z - s_2} \right)\frac{d\xi}{dz} + \left(\frac{\frac{3}{16}}{z^2} + \frac{(b - \frac{3}{16})z + a - \frac{3}{16}(s_1 + s_2)}{z(z - s_1)(z - s_2)} \right)\xi = 0, \tag{7.6}$$

where $s_i = \frac{1}{e_i}$, $i = 1, 2$.

The next reduction is obtained by the change ([88], p. 78)

$$\xi(z) = z^{1/4}\eta(z). \tag{7.7}$$

By the above change, (7.6) is transformed into

$$\frac{d^2\eta}{dz^2} + \frac{1}{2}\left(\frac{1}{z} + \frac{1}{z - s_1} + \frac{1}{z - s_2} \right)\frac{d\eta}{dz} + \frac{(4b + \frac{1}{4})z + 4a + \frac{1}{4}(s_1 + s_2)}{4z(z - s_1)(z - s_2)}\eta = 0. \tag{7.8}$$

With the change of the independent variable

$$p = z - \frac{1}{3}(s_1 + s_2) \qquad (7.9)$$

(7.8) becomes the standard algebraic form of the Lamé equation (see Subsection 2.8.4)

$$\frac{d^2\eta}{dp^2} + \frac{f'(p)}{2f(p)}\frac{d\eta}{dp} - \frac{Ap+B}{f(p)}\eta = 0, \qquad (7.10)$$

where $f(p) = 4p^3 - g_2 p - g_3$, with $g_2 = \frac{4}{3}(s_1 + s_2)^3 - s_1 s_2$, $g_3 = -\frac{4}{27}(s_1 + s_2)(s_2 - 2s_1)(s1 - 2s_2)$ and $A = -(4b + \frac{1}{4})$, $B = -(4b + \frac{1}{4})\frac{1}{3}(s_1 + s_2) - 4a - \frac{s_1 + s_2}{4}$.

Finally, with the well-known change $p = \mathcal{P}(\tau)$, we get the Weierstrass form of the Lamé equation

$$\frac{d^2\eta}{d\tau^2} - (n(n+1)A\mathcal{P}(\tau) + B)\eta = 0, \qquad (7.11)$$

$\mathcal{P}$ being the elliptic Weierstrass function.

We recall that this equation is defined in a torus Π (genus one Riemann surface) with only one singular point at the origin. We have denoted by $2w_1$, $2w_3$ the real and imaginary periods of the Weierstrass function $\mathcal{P}$ and $\mathbf{g}_1$, $\mathbf{g}_2$ their corresponding monodromies in the above equation. If $\mathbf{g}_*$ represents the monodromy around the singular point, then $\mathbf{g}_* = [\mathbf{g}_1, \mathbf{g}_2]$.

It is easy to see that $\Gamma_{\mathbf{R}}$ corresponds, by the global change $r \mapsto \tau$, to the real segment between the origin and $2w_1$ in the plane τ. Hence, the monodromy $m_\gamma = m_\sigma^2$ (see Section 7.2) is equal to $\mathbf{g}_1^2$. We recall now (Subsection 2.8.4) that the condition $\mathbf{g}_* = \mathbf{1}$ is equivalent to n being an integer.

We come back to our example. As $A = -(4b + 1/4)$ with $b > 0$, n is not an integer (the roots of the indicial equation are $-1/2 \pm 2i\sqrt{b}$) and $\mathbf{g}_1^2 \neq \mathbf{1}$, equivalent (by Proposition 7.3) to $(G_{\mathrm{loc}})^0$ is not abelian. Therefore, by Proposition 2.7 and Theorem 7.2 we have obtained the following non-integrability result.

Proposition 7.3 *Let*

$$H = \frac{1}{2}(y_1^2 + y_2^2) + \varphi(x_1) + \frac{1}{2}\alpha(x_1)x_2^2 + h.o.t.(x_2), \qquad (7.12)$$

be a Hamiltonian, where

$$\varphi(x_1) = -\frac{1}{2e_1 e_2}x_1^2(x_1 - e_1)(x_1 - e_2), \qquad \alpha(x_1) = \frac{3}{16e_1 e_2}x_1^2 + ax_1 + b,$$

(with real parameters $b > 0$, $e_1 \neq e_2$, a). Then the invariant manifolds of the Lyapounov orbits around the origin of the above Hamiltonian system must intersect transversally, and does not exist an additional global meromorphic first integral.

We note that the (global) Galois group G of the *NVE* is either of type 6 or 7 of Proposition 2.2, because by Proposition 7.2 and Lemma 7.1, the local Galois group G_{loc} is already of this type and $G_{\mathrm{loc}} \subset G$ (since $\overline{\Gamma}_{\mathrm{loc}} \subset \overline{\Gamma}$, see Section 2.2). We shall prove that $G = SL(2, \mathbf{C})$.

The relation between the Galois groups of the initial *NVE* (defined over $\overline{\Gamma}$) and equation (7.11) is given by the following lemma.

Lemma 7.2 *The identity component of the Galois groups of the* NVE *(equation (6)) and of equation (7.11) are the same (up to isomorphism).*

Proof. First, all the identity components G^0 of the Galois groups of each of the above equations (7.5), (7.6), (7.8), (7.10) and (7.11) are the same (up to isomorphism). In fact, the equivalence between (7.5) and (7.6), and between (7.8) and (7.10) is clear. On the other hand, in the change (7.7) we have introduced algebraic functions only, and these do not affect the identity component. Furthermore, by Theorem 2.5, the coverings $\mathbf{P}^1 = \overline{\Gamma} \to \mathbf{P}^1$ ($r \mapsto x$) and $\Pi \to \mathbf{P}^1$ ($\tau \mapsto p$) preserve the identity component of the Galois group. Hence, the identity component of the Galois group of equation (7.6) is the same as the identity component of the Galois group of the *ANVE* (equation (7.5)). $\qquad\square$

Now we shall compute the identity component of the Galois group of equation (7.11).

We recall that the roots of the indicial equation at the origin are $-1/2 \pm 2i\sqrt{b}$. The eigenvalues of the corresponding monodromy matrix $\mathbf{g}_*$ are not in the unit circle, and cases 1, 2 and 3 of Proposition 2.2 are not possible. As n is not an integer, the abelian case 4 (of Proposition 2.2) is also impossible. We can not have case 5, for this meta-abelian case does not appear in the Lamé equation (by Proposition 2.6). If we are in case 6, the commutator of the monodromy matrices along the periods, $\mathbf{g}_* = [\mathbf{g}_1, \mathbf{g}_2]$, has eigenvalues equal to 1. Necessarily we are in case 7, $G = G^0 = SL(2, \mathbf{C})$. Then, by the above lemma, the identity component for the *NVE* (7.6) is also $SL(2, \mathbf{C})$ and its Galois group must be $SL(2, \mathbf{C})$.

Finally, we remark that the family of (complex) Hamiltonians

$$H = -\frac{1}{2}(y_1^2 + y_2^2) + \varphi(x_1) + \frac{1}{2}\alpha(x_1)x_2^2 + h.o.t.(x_2), \qquad (7.13)$$

is obtained from the family defined by formulas (7.2), (7.3) by the symplectic (canonical) transformation $y \mapsto iy$, $t \mapsto it$. Hence, the above family and our initial family represent the same Hamiltonian system, and Proposition 7.3 is true for both families (it is implied that in the family (7.13) the phase space is given by the coordinates (x_1, x_2, iy_1, iy_2), with x_1, x_2, iy_1, iy_2 real).

Chapter 8

Complementary Results and Conjectures

In this last chapter we give some additional results, and we also formulate some conjectures that open new lines of research which are in progress. We will not enter into the computational details and this chapter is written in a more informal style than the rest of this monograph.

It seems clear that the linear equation that appears most frequently in applications is the hypergeometric (or Riemann) equation and its confluent versions (as Bessel equation). In Section 8.1 we will again apply the theorem of Kimura (Subsection 2.8.1) and our results of Chapter 4 to two systems: the Spring-Pendulum and a generalization of the homogeneous potentials.

Section 8.2 is devoted to two conjectures on the chaotic dynamics of non-integrable systems.

In Section 8.3 we propose a generalization of the theory to higher order variational equations. This opens a new line of research and one more conjecture is formulated. As an example we apply this to the Hénon-Heiles system for which the first order approach is not conclusive (Section 6.3, Example 1). The section ends with some comments about a (possible) connection with the so-called Painlevé test and the Poincaré-Arnold-Mellnikov integral.

8.1 Two additional applications

The Spring-Pendulum system is a Hamiltonian system which, in suitable coordinates, is defined by the Hamiltonian (see [23] and references therein)

$$H = \frac{1}{2}(y_1^2 + \frac{y_2^2}{x_1^2}) + \frac{1}{2}x_1^2 + x_1(\lambda - 1 - \lambda\cos x_2). \qquad (8.1)$$

J. J. Morales Ruiz, *Differential Galois Theory and Non-Integrability of Hamiltonian*, Systems, Modern Birkhäuser Classics,
DOI: 10.1007/978-3-0348-0723-4_8, © Springer Basel 1999

We can expand the above function in powers of the variable x_2 (as in Chapter 6) and get

$$H = \frac{1}{2}\left(y_1^2 + \frac{y_2^2}{x_1^2}\right) + \frac{1}{2}x_1^2 - x_1 + \frac{\lambda}{2}x_1 x_2^2 + O(x_2^3).$$

We remark that for the physical model $0 \le \lambda \le 1$.

Then the plane $x_2 = y_2 = 0$ is invariant and, in the reference above, the authors studied the NVE along the integral curves contained in this plane. In fact they made a covering $z := \frac{1}{2}x_1 \mapsto t$ and obtained an algebraic form of the NVE, $ANVE$, as a hypergeometric (or Riemann) equation

$$\frac{d^2\xi}{dz^2} + \left(\frac{5/2}{z} + \frac{1/2}{z-1}\right)\frac{d\xi}{dz} + \left(\frac{\lambda/2}{z^2} + \frac{\lambda/2}{z(1-z)}\right)\xi = 0. \qquad (8.2)$$

We recall that NVE and equation (8.2) have the same identity component of the Galois group (up to isomorphism) (Theorem 2.5).

Then it is proved in reference [23] that a necessary condition for integrability is

$$\lambda = \frac{1}{8}(9 - q^2),$$

q being a rational number.

In order to apply Kimura's theorem (Theorem 2.6) we compute the difference of exponents. They are $\hat{\lambda} = \frac{1}{2}\sqrt{9 - 8\lambda}$, $\hat{\mu} = \frac{1}{2}$ and $\hat{\nu} = 2$. Then by Kimura's theorem (and our Theorems 2.5 and 4.1) we get the following result (we leave the details as an exercise).

Proposition 8.1 *The Hamiltonian system defined by (8.1) is non-integrable with a meromorphic first integral, except if $\lambda = \frac{1}{2}(2 - p(p+1))$, p being an integer.*

We observe that for λ in the physical region $0 \le \lambda \le 1$, we get non-integrability if $0 < \lambda < 1$. The cases $\lambda = 0, 1$ are separable. So, the integrability problem for λ in the physical domain is completely solved.

As a second example we generalize our non-integrability result on the model of homogeneous potentials of Section 4.1 to Hamiltonian systems of the type

$$H = T + V, \qquad (8.3)$$

V being, as above, a homogeneous function of degree k of the positions $x_1, \ldots, x_n$, but T is now a *homogeneous function* of degree m of the moments $y_1, \ldots, y_n$ (in Section 4.1 we studied a particular case with $m = 2$).

Using similar arguments to those of the Section 4.1, Yoshida showed that, under some restrictions on the homogeneous functions T and V, it is also possible to obtain a direct sum of hypergeometric equations as a pull-back of a finite covering of the *NVE* along an algebraic curve contained in some invariant plane. As in Subsection 5.1.2, we call it the algebraic normal variational equation, $ANVE = ANVE_1 \oplus ANVE_2 \oplus \cdots \oplus ANVE_{n-1}$, and each $ANVE_i$ is a hypergeometric equation (2.11) with parameters

$$a + b = 1/m - 1/k, \; ab = -\frac{\lambda_i}{mk}, \; c = 1 - \frac{1}{k};$$

here the Yoshida coefficients, λ_i, are the product of the eigenvalues of the Hessian of T and of the eigenvalues of the Hessian of V (obtained by simultaneous diagonalization of both quadratic forms). As usual we do not consider the parameter λ_n corresponding to the tangential variational equation (see [113] for details).

Then by the theorem of Kimura (and Theorems 2.5, 4.1 and Corollary 4.1) we obtain a generalization of Theorem 5.1. As the explicit list of the values of the parameters (m, k, λ_i) compatible with integrability is very long in this case, it is better to check Kimura's theorem for the particular system under study. Then we give the result in a indirect way.

Proposition 8.2 *Assume that Hamiltonian system given by the Hamiltonian (8.3) is completely integrable with meromorphic first integrals, then for each $i = 1, 2, \ldots, n - 1$, the values (m, k, λ_i) are such that the corresponding hypergeometric equation must satisfy one of the conditions in (i) or (ii) of the theorem of Kimura (Theorem 2.6).*

Theorem 5.1 is a corollary of the above result.

In his paper Yoshida considers the two degrees of freedom Hamiltonian

$$H = F(y_1, y_2) + F(x_1, x_2),$$

with

$$F(z_1, z_2) := \frac{1}{4}(z_1^4 + z_2^4) + \frac{e}{2}z_1^2 z_2^2.$$

This Hamiltonian has two invariant planes with an *ANVE* of hypergeometric type and with Yoshida coefficients

$$e^2$$

and

$$\left(\frac{3 - e}{1 + e}\right)^2,$$

respectively (see [113] for details).

We leave it as an exercise to apply the above proposition to this system and to obtain necessary conditions on the parameter e in order to have integrability.

The possibility of applying our results to Hamiltonian systems given by (8.3) was suggested to the author by Haruo Yoshida.

8.2 A conjecture about the dynamic

Except for Chapter 7 and for the last section of Chapter 6, this monograph has been devoted to algebraic non-integrability criteria for Hamiltonian systems. So, the main theorems of Chapter 4 are of algebraic character, and we do not know the dynamic of the system in the general situation. Question 2 of Section 6.3 formulates the general problem. However, in the particular hyperbolic context of Chapter 7, we were able to prove that the algebraic obstruction to integrability, discussed in Chapter 4, is equivalent to the dynamical one given by Lerman's theorem, and chaotic dynamics is given by the splitting of real invariant manifolds. We think that this is not a pathological behaviour, typical of this kind of system only, but that it is also true in more general situations. So we formulate the following conjecture.

Conjecture 1 *If a complex analytical Hamiltonian system is not completely integrable with meromorphic first integrals, then it has a chaotic behaviour in some part of the complex phase space.*

It is clear that the formulation of the above conjecture is necessarily ambiguous due to the context in which it is formulated. A more precise conjecture, closer to the results of Chapter 7, can be formulated as follows.

Conjecture 2 *Assume that a complex analytical Hamiltonian system contains a completely integrable subsystem with meromorphic first integrals and that the variational equation along an integral curve, contained in this subsystem, is non-integrable in the sense of the differential Galois theory. Then the complex Hamiltonian system has a chaotic behaviour in some neighborhood of the above integral curve.*

Conjecture 2 is being studied by Josep M. Peris and the author.

8.3 Higher-order variational equations

By commodity of notation, as in other chapters of this monograph, in this section we identify the Riemann surface Γ, defined by an integral curve, with the integral curve $i(\Gamma)$. So, we say, for instance, that T_Γ is the tangent bundle of the manifold M restricted to the integral curve Γ, etc. . . .

8.3.1 A conjecture

So far in this monograph we have considered the variational equation of order one of a Hamiltonian system, X_H, along a particular integral curve Γ. This variational equation, VE, gives us the linear part of the flow of X_H along Γ, but we can also consider the quadratic, cubic, etc... contributions to the flow along Γ (i.e., the rest of the jet along Γ). These higher order terms are given by the higher order variational equations along Γ and it is well known that they can be solved by the method of variation of constants in a recurrent way, once we know the solution of the order one variational equation.

The problem now is to study the possible extension of our first order non-integrability criterion to higher order (this problem was proposed by Carles Simó to the author some years ago). We will not solve this problem in a complete way here, but we try to gain some insight into its solution.

We start by fixing the terminology and notation. Let

$$\dot{z} = X(z), \qquad (8.4)$$

be a non-linear differential equation defined on a manifold M, $z \in M$.

Let $\phi(z, t)$ be the flow defined by the above equation and let $z(t) = \phi(z_0, t)$ be the function that represents the integral curve Γ of the field X in the temporal parametrization, such that $z_0 = \phi(z_0, t_0)$. Then, we denote by $\phi^{(1)}$, $\phi^{(2)}, \ldots,$ $\phi^{(k)}, \ldots,$ the derivatives of ϕ with respect to z at the point (z_0, t) of orders $1, 2, \ldots, k, \ldots,$ respectively (in fact, the functions $\phi^{(k)} = \phi^{(k)}(t)$ are defined over the universal covering of Γ).

For a given k, the set of functions $\{\phi^{(1)}, \phi^{(2)}, \ldots, \phi^{(k)}\}$ satisfy the variational equation up to order k, $(VE)_k$. In particular, the variational equation $(VE)_1$ is the first order variational equation written as VE in this monograph. We observe that to compute the variational equation $(VE)_k$, it is convenient to write equation (8.4) as

$$\frac{\partial \phi}{\partial t} = X(\phi(z, t)). \qquad (8.5)$$

Then we compute the successive derivatives of equation (8.5), up to order k, with respect to the variable z and we substitute $\phi(z_0, t) = z(t)$. This set of "prolongated" equations is $(VE)_k$. It is clear that $(VE)_{k-1}$ is contained in $(VE)_k$. Furthermore, $\phi^{(1)}$ is the solution of $(VE)_1$ with initial conditions $\phi^{(1)}(t_0) = I$ (identity), given by a fundamental system of solutions of $(VE)_1$), and for the other functions $\phi^{(s)}$, $s = 2, \ldots, k$, we take as initial conditions $\phi^{(s)}(t_0) = 0$ (because these functions are the coefficients of the Taylor expansion of $\phi(z, t)$ at the point (z_0, t)).

Let now K be the differential field of coefficients of $(VE)_1$ as in Chapter 4. We recall that K is always a field of meromorphic functions over a Riemann surface, but the Riemann surface considered depends on the context, $\Gamma, \overline{\Gamma}$, etc... (see Sections 4.1 and 4.2). Then for a given k, to the variational equation up to order k, $(VE)_k$, we associate the chain of extensions of differential fields

$$K := L_0 \subset L_1 \subset L_2 \subset \cdots \subset L_{k-1} \subset L_k,$$

with $L_1 = K(\phi^{(1)})$, $L_2 = L_1(\phi^{(2)}) = K(\phi^{(1)}, \phi^{(2)}), \ldots,$ $L_k = K(\phi^{(1)}, \ldots, \phi^{(k)})$.
We know that the first extension of the above chain, $K \subset L_1$, is a Picard-Vessiot
extension. The other individual extensions $L_s \subset L_{s+1}$ are also Picard-Vessiot
extensions because they are obtained by the method of variation of constants,
i.e., by quadratures, and the Galois group $\mathrm{Gal}_{L_{s-1}}(L_s)$ is a vector group (i.e.,
isomorphic to the additive group $(\mathbf{C}^r, +)$, for some r, see Section 2.2). From this
it is not evident that the total extension $K \subset L_k$ is a Picard-Vessiot extension
(a composition of Picard-Vessiot extensions is not, in general, a Picard-Vessiot
extension!), but this is indeed the case.

Proposition 8.3 *For each* $k \geq 1$, *the extension* $K \subset L_k$ *is a Picard-Vessiot*
extension.

The key point in this proposition is that the variational equation up to
order k, $(VE)_k$, is a linear differential equation *with the same coefficient field*
K as the first order variational equation $(VE)_1$.

We do not give here a general proof of the proposition above, but we
prove it for $(VE)_2$ and $(VE)_3$, since they are the only cases necessary for the
application discussed below.

The second variational equation, $(VE)_2$ is given by

$$\dot{\phi}^{(1)} = X^{(1)}\phi^{(1)}, \tag{8.6}$$
$$\dot{\phi}^{(2)} = X^{(1)}\phi^{(2)} + X^{(2)}(\phi^{(1)}, \phi^{(1)}), \tag{8.7}$$

where for simplicity, according to our prescription, we write $X^{(1)}$ and $X^{(2)}$ for
$X^{(1)}(z(t))$ and $X^{(1)}(z(t))$, respectively.

Now it is convenient to introduce some notation. Let E be a vector space,
given two linear applications $f : S^r E \longrightarrow E$ and $g : S^s E \longrightarrow E$ ($S^r E$ and $S^s E$
are symmetric tensor powers), then we define the symmetric product of f and
g as the linear map

$$f \bullet g : S^{r+s} E \longrightarrow S^2 E,$$

$(f \bullet g)(u \bullet v) = f(u) \bullet g(v)$, with $\bullet$ being the symmetric product (i.e., the product
in the symmetric algebra). As usual, if $f = g$, we write $f \bullet g = S^2 f$.

Consider the linear differential equation

$$S^2\dot{\phi}^{(1)} = BS^2\phi^{(1)}, \tag{8.8}$$
$$\dot{\phi}^{(2)} = X^{(1)}\phi^{(2)} + X^{(2)}S^2\phi^{(1)}, \tag{8.9}$$

where equation (8.8) corresponds to the construction $S^2\nabla_1$ of the connection
∇_1 of the first order variational equation (8.6) (i.e., we compute the derivative
of $\phi^{(1)} \bullet \phi^{(1)}$ by the Leibniz rule and then we apply $(VE)_1$). Matrix B depends
on $X^{(1)}$ and its coefficients belong to the differential field K. Equation (8.9) is

equation (8.7) written in tensorial form. We denote the connection of the above system of equations, (8.8)+(8.9), as ∇_2. Then we have the exact sequence of connections (see Section 2.3, Example 3)

$$0 \longrightarrow (V_1, \nabla_1) \longrightarrow (V_1 \oplus V_2, \nabla_2) \longrightarrow (V_2, S^2 \nabla_1) \longrightarrow 0,$$

and the linear equation (8.8)+(8.9) is solved from the solution of $\phi^{(1)}$ of $(VE)_1$ by the method of variation of constants. From this it is easy to see that the extension $K \subset L_2$ is the Picard-Vessiot extension of the connection ∇_2.

In a similar, but in a more involved way, we can construct a linear connection ∇_3 for the third order variational equation $(VE)_3$. The associated linear equation is given by the system of equations

$$
\begin{align}
S^3 \dot{\phi}^{(1)} &= C S^3 \phi^{(1)}, \tag{8.10} \\
\phi^{(1)} \bullet \dot{\phi}^{(2)} &= B(\phi^{(1)} \bullet \phi^{(2)}) + D S^3 \phi^{(1)}, \tag{8.11} \\
\dot{\phi}^{(3)} &= X^{(1)} \phi^{(3)} + X^{(2)}(\phi^{(1)} \bullet \phi^{(2)}) + X^{(3)} S^3 \phi_1, \tag{8.12}
\end{align}
$$

where the matrix B is the same as above and the matrices C and D depend in a polynomial way on (the matrix elements of) $X^{(1)}$ and $X^{(2)}$, respectively, and they have their coefficients in K. We remark that the map $X^{(2)}(\phi^{(1)} \bullet \phi^{(2)})$, considered as a trilinear symmetric map, is explicitly given by

$$(\phi^{(1)} \bullet \phi^{(2)})(\xi_1, \xi_2, \xi_3) = (\phi^{(1)} \xi_1, \phi^{(2)}(\xi_2, \xi_3)) + \text{ circular permutations of } 1, 2, 3$$

(this is the "crossed" term which appears in the chain rule when we compute the third derivative).

If ∇_3 is the connection associated to (8.10)+(8.11)+(8.12), then as above we have the exact sequence

$$0 \longrightarrow (V_1 \oplus V_2, \nabla_2) \longrightarrow (V_1 \oplus V_2 \oplus V_3, \nabla_3) \longrightarrow (V_3, S^3 \nabla_1) \longrightarrow 0.$$

Then the extension $K \subset L_3$ is the Picard-Vessiot extension of the connection ∇_3.

We notice that from the above proposition follows

Corollary 8.1 *The extension L_k/L_1 is a Picard-Vessiot extension and $\mathrm{Gal}_{L_1}(L_k)$ is a connected linear algebraic group (Zariski topology).*

Proof. The fact that L_k/L_1 is a Picard-Vessiot extension follows from the differential Galois theory (see Theorem 2.2). As each extension L_s/L_{s-1} is a purely transcendental one (this is equivalent to the connectedness of $\mathrm{Gal}_{L_{s-1}}(L_s)$), then the total extension L_k/L_1 is also transcendental and its Galois group is connected. $\qquad\square$

In particular, the group $\mathrm{Gal}_{L_1}(L_k)$ is contained in the identity component $(G_k)^0$ of $G_k := \mathrm{Gal}_K(L_k)$.

If now $X = X_H$ is an integrable Hamiltonian system, under the same assumptions of the results of Section 4.2 (Theorems 4.1, 4.2, 4.3 or Corollaries 4.3, 4.4, depending on the context), a natural extension of our philosophy is that the non-linear abelian structure given by the integrability of X_H must be projected on the variational equation at any order. Hence we state the following conjecture.

Conjecture 3 *Assume that the Hamiltonian system X_H is integrable, then the identity component of the Galois group of the extension $K \subset L_k$ is abelian for any $k \geq 1$.*

By the rigidity of the complex analytical structures, it is natural to ask about the existence of a *sufficient* condition of integrability, under suitable assumptions. Assume that the Hamiltonian system is not integrable and that the integral curve Γ, along which we compute the variational equations, is contained in an integrable subsystem of X_H. Then we formulate the following question.

Question *Is the identity component of the Galois group of the variational equation $(VE)_k$ not abelian for some k?*

We remark that if the variational equation of order one is integrable (in the differential Galois sense), then all the higher order variational equations are also solvable, because they are solved by the method of variations of constants in a recurrent way, once we know the solution of the first order variational equation.

The proof of Conjecture 3 is in progress in a joint work of Jean-Pierre Ramis, Carles Simó and the author [80].

8.3.2   An application

Now we apply the above conjecture to the Hénon-Heiles potential of Section 6.3 with $d/c = 2$. We recall that this is the only case in the family of the Hénon-Heiles Hamiltonians whose integrability remains open.

If we assume that Conjecture 3 is true, we can prove the non-integrability of this system. We give the main ideas underlying the proof.

The Hamiltonian system can be written as

$$H = \frac{1}{2}(y_1^2 + y_2^2) + \frac{1}{2}x_1^2 + \frac{1}{3}x_1^3 + \frac{1}{2}x_1 x_2^2. \tag{8.13}$$

The proof is based on the following two lemmas. Let G_k be the Galois group of the variational equation up to order k, $(VE)_k$, along one of the family of elliptic integral curves Γ, parametrized by the energy, and contained in the invariant plane $x_2 = y_2 = 0$ (as in Chapter 6). We have shown in Chapter 6 that Γ is a punctured torus and that the first order variational equation $(VE)_1$ splits into a direct sum of the tangential variational equation and the normal variational equation (because Γ is contained in an invariant plane). Both of them are of Lamé type. We write that as

$$\nabla_1 = (\nabla_1)_T \oplus (\nabla_1)_N.$$

Then it is not difficult to verify that the Galois group $G_1 = \text{Gal}(\nabla_1)$ is represented by unipotent triangular matrices of the type

$$\begin{pmatrix} 1 & 0 & 0 & 0 \\ \alpha & 1 & 0 & 0 \\ 0 & 0 & 1 & 0 \\ 0 & 0 & \beta & 1 \end{pmatrix};$$

because for both, the tangential and the normal first order variational equations, we are in the case of solvability given by Lamé, where one non-trivial solution of the Lamé equation is an elliptic function, i.e., it belongs to the coefficient field K (see Subsection 2.8.4, or for more information [42, 88, 109]).

Notice that the Galois group G_1 is abelian, for this reason it is not possible to obtain a non-integrability criterion by an analysis up to first order.

As the general solution of $(VE)_1$ is obtained by two quadratures, then the extension L_1/K of the first order variational equation is transcendental (this is also easily proved by observing that G_1 is isomorphic to the vector group $(\mathbf{C}^2, +)$), and as the extensions L_k/L_1 are also transcendental, we have the following lemma.

Lemma 8.1 *The Galois group G_k of the variational equation up to order k of the Hamiltonian system (8.13), along the elliptic integral curves contained in the invariant plane $x_2 = y_2 = 0$, is connected, i.e., $G_k = (G_k)^0$.*

Now the coefficients of the order k variational equations are elliptic functions holomorphic over a punctured torus, and this singularity is a regular singular one of this variational equation (because their solutions are obtained by the method of variations of constants from the solution of the first order variational equation). Then we arrive at the same necessary and sufficient condition for abelianess of the monodromy (and Galois) group for the Lamé equation (see Subsection 2.8.4): $\mathbf{g}_* := [\mathbf{g}_1, \mathbf{g}_2]$ around the singular point must be the identity.

Lemma 8.2 *The Galois group G_k of the variational equation of order k, $(VE)_k$, of the Hamiltonian system given by (8.13) is abelian, if and only if, the local monodromy $\mathbf{g}_*$ of the variational equation $(VE)_k$, around the singular point of the coefficients is the identity.*

Theorem 8.1 *The Hamiltonian system defined by the function* (8.13) *has no additional meromorphic integral independent of the Hamiltonian, provided Conjecture 3 is true.*

Sketch of Proof. By Conjecture 3 we need to find some k such that $(G_k)^0$ is not abelian. But by Lemma 8.1 this is equivalent to saying that G_k is not abelian, which means equivalent, by Lemma 8.2, to g_* different from the identity for some k. Then we start with the solution of the first order variational equation, and we solve the second order variational equation by the method of variation of constants, then the third order variational equation by the same method and so on. ... If in some step of this process we get a logarithm around the singular point (i.e., a residue different from zero in the Laurent series of some integrand) then $\mathbf{g}_*$ is not the identity for the corresponding variational equation, and G^0 is not abelian. A long but straightforward computation gives us a logarithm for $k = 3$.   □

   The explicit computation in the above proof has been done by Carles Simó. In the above cited forthcoming paper by Jean-Pierre Ramis, Carles Simó and the author [80], the details of this computation will be explained.

   We finish with two informal remarks. We note the analogy between the method used in the above proof and the so-called Painlevé test: the existence of a logarithmic term in the jet is an obstruction to integrability (see [37]).

   Also, there exists an analogy with the Poincaré-Arnold-Mellnikov integral method (see [37], where a connection with the Painlevé test is studied). We recall that this method is essentially the method of variation of constants applied to the variational equation *with respect to parameters* of a perturbed Hamiltonian system, the unperturbed system being integrable. Then the homogeneous part of the variational equation is integrable. In fact, by the main theorems of Chapter 4, the identity component of its Galois group is abelian, because the homogeneous part of the variational equation with respect to the parameters coincides with the first order variational equation of the unperturbed system with respect to the initial conditions. Furthermore, it is possible to check that the Poincaré-Arnold-Mellnikov integral is an object related to a tensorial construction of the variational equations (concretely to an exterior product). We do not pursue in this direction, but we claim for the formulation of a theory that contains as particular cases the theory of non-integrability with respect to the initial condition (as formulated along this monograph) and the Poincaré-Arnold-Mellnikov theory.

Appendix A

Meromorphic Bundles

In this appendix we give the precise definition of meromorphic vector bundles and their trivializations in the symplectic case. We follow the paper [77].

Let X be a Riemann surface. We denote by $\mathcal{O}_X$ and by $\mathcal{M}_X$ the sheaves of holomorphic and meromorphic functions on X. The sheaf of holomorphic sections $\underline{V}$ of a *holomorphic vector bundle* V of rank n is a sheaf of $\mathcal{O}_X$-modules that is locally isomorphic to $\mathcal{O}_X^n$. A holomorphic vector bundle of rank n on X is also interpreted as an element of the non-abelian cohomology set $H^1(X; GL(n; \mathcal{O}_X))$.

Let $G \subset GL(n; \mathbf{C})$ be an algebraic subgroup (defined on the field of complex numbers). We set $\mathbf{G}^{an} = G_{\mathcal{O}_X} \subset GL(n; \mathcal{O}_X)$ and $\mathbf{G}^{me} = G_{\mathcal{M}_X} \subset GL(n; \mathcal{M}_X)$. We say that a holomorphic bundle on X admits G as structure group if it is defined by an element of $H^1(X; \mathbf{G}^{an})$.

We need *meromorphic vector bundles.* By definition, the sheaf of meromorphic sections of a meromorphic vector bundle of rank n is a sheaf of $\mathcal{M}_X$-modules that is locally isomorphic to $\mathcal{M}_X^n$. A meromorphic vector bundle of rank n on X is also interpreted as an element of the non-abelian cohomology set $H^1(X; GL(n; \mathcal{M}_X))$. If this element "belongs" to $H^1(X; \mathbf{G}^{me})$, we say that the meromorphic vector bundle admits G as structure group. There exists an equivalent definition for a meromorphic vector bundle on a Riemann surface X, due to Deligne ([30], 1.14 p. 52). Such a bundle is an equivalence class of *holomorphic extensions* of holomorphic bundles defined on X minus a discrete subset Σ: locally, if z is a uniformizing variable vanishing on Σ, then two extensions V_1 and V_2 of V are equivalent if the corresponding sheaves of holomorphic sections satisfy

$$z^n \underline{V}_1 \subset \underline{V}_2 \subset z^{-n} \underline{V}_1 \subset i^* \underline{V}$$

$(i : X - \Sigma \to X$ being the natural inclusion).

J. J. Morales Ruiz, *Differential Galois Theory and Non-Integrability*
of Hamiltonian, Systems, Modern Birkhäuser Classics,
DOI: 10.1007/978-3-0348-0723-4, © Springer Basel 1999

The following result says that every meromorphic vector bundle on a Riemann surface comes from a holomorphic vector bundle.

Lemma A.1 *Let X be a Riemann surface. Let $G \subset GL(n; \mathbf{C})$ be an algebraic subgroup defined on the field of complex numbers. Then the natural map*

$$H^1(X; \mathbf{G}^{an}) \to H^1(X; \mathbf{G}^{me})$$

is surjective.

The proof is easy: the set of poles of a section of $\mathbf{G}^{me}$ is discrete.

Proposition A.1 *Any meromorphic vector bundle on a Riemann surface X is trivial.*

Proof. Let Vme be a meromophic vector bundle on X. It comes from a holomorphic vector bundle V^{an}. If X is an open Riemann surface, then V^{an} is trivial ([36], Theorem 30.4). If X is a compact connected Riemann surface, then V^{an} comes from an *algebraic* vector bundle V on the non-singular projective curve X. We denote by k_X the field of rational (or meromorphic) functions on X. The field of rational sections of the algebraic bundle V is a rank n vector space on k_X ([36], Corollary 29.17), therefore Vme is a trivial meromorphic bundle. $\square$

In fact we need some similar, but more precise, results involving vector bundles with the symplectic group as structure group. We will give them below.

If now X is a *singular* complex analytic curve, we can also define holomorphic vector bundles and meromorphic vector bundles on X along the same lines. If $\pi : \tilde{X} \to X$ is a desingularization map (i.e., if $\tilde{X}$ is a Riemann surface and π a proper analytic map, which is a finite covering over a discrete subset of X), then it induces an isomorphism π^* between the sheaves of meromorphic functions $\mathcal{M}_X$ and $\mathcal{M}_{\tilde{X}}$, and therefore an isomorphism π^* between the meromorphic vector bundles on X and on $\tilde{X}$.

Theorem A.1 (Grauert Theorem) *Let X be a complex connected, non-compact, Riemann surface. Let $\mathcal{F} = (Y, p, X)$ be a locally trivial vector (resp., principal) holomorphic fibre bundle on X with a connected complex Lie group G as structure group. Then $\mathcal{F}$ is holomorphically trivial.*

Sketch of Proof. For completeness we recall here the proof. We denote by $\mathbf{G}^{an}$ (resp., $\mathbf{G}^c$) the sheaf of holomorphic (resp., continuous) functions on X with values in G. The open Riemann surface X is homotopically equivalent (by retraction) to a finite one-dimensional complex. On such a complex, a G-fibre bundle is topologically trivial because G is connected. Therefore the fibre bundle $\mathcal{F}$ is topologically trivial. An open Riemann surface is a Stein manifold. On such a manifold the topological and the analytic classifications of fibre bundles with a complex Lie group as structure group coincide: the natural map

$$H^1(X; \mathbf{G}^{an}) \to H^1(X; \mathbf{G}^c)$$

is a bijection [39, 17]. Therefore $\mathcal{F}$ is holomorphically trivial. ☐

A complete and detailed proof is given in [92], Chapter 2.

We apply the above theorem to the symplectic group. An element of $Sp(2n, \mathbf{C})$ is a product of at most $4n - 2$ symplectic transvections [31] (we can also use a homeomorphism between $Sp(2n, \mathbf{C})$ and the product of $SU(n) \times$ *vector space* and the connectedness of $SU(n)$). Hence

Lemma A.2 *The topological group $Sp(2n, \mathbf{C})$ is connected.*

Corollary A.1 *Let X be a complex connected, non-compact, Riemann surface. Let $\mathcal{F} = (Y, p, X)$ be a locally trivial vector (resp., principal) holomorphic fibre bundle on X with $Sp(2n, \mathbf{C})$ as structure group. Then $\mathcal{F}$ is holomorphically trivial.*

We study now the case of bundles on compact Riemann surfaces.

Proposition A.2 *Let X be a complex connected compact Riemann surface. Let $\mathcal{F} = (Y, p, X)$ be a locally trivial holomorphic vector (resp., principal) fibre bundle on X with structure group $G = Sp(2n, \mathbf{C})$. Then $\mathcal{F}$ is meromorphically trivial.*

Proof. The proof is based on the "GAGA" paper of Serre. The compact Riemann surface X is also a complex algebraic (projective) curve. We denote by $\mathbf{G}$ the sheaf of regular maps from X to the algebraic complex group G. We have a natural map

$$L : H^1(X; \mathbf{G}) \to H^1(X; \mathbf{G}^{an}).$$

The symplectic group $G = Sp(2n, \mathbf{C})$ satisfies condition (R) of Serre ([91], p. 33): there exists a rational section

$$GL_{2n}(\mathbf{C})/G \to GL_{2n}(\mathbf{C})$$

([91], Example c) p. 34). Therefore, we can apply Proposition 20 of Serre ([91], p. 33): the map L is a bijection. Using an algebraic trivialisation of the algebraic bundle corresponding to $\mathcal{F}$ on a convenient affine subset of the curve X, we get the result. ☐

Now we shall apply the above to the bundles we are interested in.

Let M' be a connected complex analytic manifold of complex dimension $2n$. Let Ω be a closed meromorphic form of degree two on M'. Let $M_\infty \subset M'$ be a closed analytic hypersurface (i.e., analytic subset of pure complex codimension one) of M'. We set $M = M' - M_\infty$ and we suppose that Ω is *holomorphic* and *non-degenerated* over M. Then (M, Ω) is a complex symplectic manifold. We denote by $T\,M'$ (resp., $T^*\,M'$) the tangent (resp. cotangent) bundle of M'. It is a holomorphic bundle but we will use only its structure of meromorphic bundle. Then, as we noticed before, the form Ω induces a generalized musical isomorphism between the *meromorphic* bundle $T\,M'$ and the *meromorphic* bundle $T^*\,M'$: if X_1 is a meromorphic vector field on an open set $U \subset M'$, then, for every meromorphic vector field X on U, $\Omega(X_1, X)$ is a meromorphic function on U, and

$$X \to \Omega(X_1, X)$$

is a k_U-linear isomorphism between the k_U-vector spaces of meromorphic sections of $T\,M'$ and $T^*\,M'$ on U. We denoted by k_U the field of meromorphic functions on U.

Let H be a meromorphic Hamiltonian function over the manifold M'. Let $X_H = \sharp dH$ be the corresponding Hamiltonian field. It is meromorphic over M' and its restriction to M is *holomorphic*. Let $i(\Gamma)$ be a connected non-equilibrium integral curve of X_H over M. Let $\underline{\Gamma}'$ be as before a (perhaps) singular curve that is the union of $i(\Gamma)$ and of a discrete subset of equilibrium points and points at infinity. Let $\overline{\Gamma}'$ be a desingularization of $\underline{\Gamma}'$. Let $f_1, \ldots, f_m$ be an involutive set of first integrals ($H = f_1$) that are *meromorphic* on M'. We suppose that they are *holomorphic* and *independent* at some point of the phase curve $i(\Gamma)$. Then the system $df_1 = 0, \ldots, df_m = 0$ defines a *meromorphic* subbundle E of $T_{M'}$ of rank $2n - m$. The *meromorphic* vector fields $X_1 = \sharp df_1, \ldots, X_m = \sharp df_m$ generate a rank m meromorphic subbundle F of E. Then $F^\perp$ is a meromorphic subbundle. As in [8] we get a structure of symplectic meromorphic bundle on the meromorphic bundle $N = (F^\perp/F)$ over Γ' (we have only to replace holomorphic bundles by meromorphic bundles in [8]).

Finally, as in [8], we get a normal variational connection on the symplectic bundle $N = (F^\perp/F)$ over Γ'. Here the bundle and the connection are meromorphic. The bundle N is symplectically meromorphically trivializable, therefore this normal variational connection can be interpreted as a meromophic differential equation over Γ' (the NVE).

Appendix B

Galois Groups and Finite Coverings

In this appendix we prove that the identity component of the differential Galois group, of a meromorphic connection over a Riemann surface, does not change if we take inverse images by a finite ramified covering. It is an analytic version of an algebraic result of N. Katz. We follow [77].

Proposition B.1 *Let ∇ be a germ of a meromorphic linear connection at the origin of $\mathbf{C}$. We denote by K the differential field of germs of meromorphic functions, by $G = \mathrm{Gal}_K(\nabla)$ the differential Galois group of ∇ and by G^0 its identity component.*

Let H be the subgroup of G topologically generated (in the Zariski sense) by the exponential torus and all the Stokes multipliers of ∇.

We denote by $m \in G$ the actual monodromy of ∇, by M the Zariski closure in G of the subgroup generated by m, by M^0 the identity component of M and by H_1 the subgroup of G generated by H and M^0.

(i) *The subgroups H and H_1 are Zariski closed, connected, and invariant under the adjoint action of m.*

(ii) *The group G is topologically generated by H and m.*

(iii) *The group G is algebraically generated by H_1 and m, and $G^0 = H_1$*

(iv) *The image of m in $G \triangleleft G^0$ generates this finite group.*

Sketch of the Proof. The actual monodromy m and the formal monodromy $\hat{m}$ are equal up to multiplication by a product of Stokes multipliers [69, 16]. The exponential torus is (globally) invariant by the adjoint action of the formal monodromy $\hat{m}$. Then our claims follow easily from the density theorem of Ramis, Theorem 2.3, using some elementary results about linear algebraic groups [45]. □

J. J. Morales Ruiz, *Differential Galois Theory and Non-Integrability of Hamiltonian*, Systems, Modern Birkhäuser Classics,
DOI: 10.1007/978-3-0348-0723-4, © Springer Basel 1999

Lemma B.1 *Let $\nu \in \mathbf{N}^*$. Let ∇ be a germ of a meromorphic connection at the origin of the x plane, $\mathbf{C}$. We set $x = f(t) = t^\nu$. We denote by $(X, 0)$ (respectively $(X', 0)$) the germ at the origin of the x (respectively t) plane. We denote by K (resp., K') the differential field of germs of meromorphic functions on X (resp., X'). We have a natural injective homomorphism of differential Galois groups*

$$\mathrm{Gal}_{K'}(f^*\nabla) \to \mathrm{Gal}_K(\nabla),$$

which induces an isomorphism between their Lie algebras.

Proof. We set $G = \mathrm{Gal}_K(\nabla)$ and $G' = \mathrm{Gal}_{K'}(f^*\nabla)$. The field inclusion $K \subset K'$ induces a natural map

$$\varphi : G' \to G.$$

This map is clearly injective. Let $m \in G$ be the actual monodromy of ∇. Then the actual monodromy of $\nabla' = f^*\nabla$ is $m' = m^\nu$.

The connections ∇ and ∇' have the same exponential torus and the same Stokes multipliers (more precisely the map φ induces isomorphisms). We use the notations of Proposition B.1 for ∇, and similar notations for ∇'. We have clearly $H = H'$, $M^0 = M'^0$, therefore $G^0 = H_1 = H'_1 = G'^0$.   □

Theorem B.1 *Let X be a Riemann surface. We denote by K its field of meromorphic functions. We choose a differential ∂ on K. Let $S = \{a_i\}_{i \in I} \subset X$ be a discrete subset. Let $x_0 \in X - S$. For each point $a_i \in S$, we choose a germ d_i of real half line starting at a_i, and drawn on the complex line tangent to X at a_i. We denote by $\tilde{\mathcal{M}}$ the field of meromorphic functions on the universal covering $(\tilde{X}, x_0)$ of X pointed at x_0. We identify the field K with a subfield of $\tilde{\mathcal{M}}$. For $i \in I$, we denote by $\mathcal{M}_i$ the field of germs, at a_i, of meromorphic functions (i.e., of germs of functions meromorphic on a germ of an open sector at a_i bisected by d_i). We identify the field K_i of germs, at a_i, of meromorphic functions with a subfield of $\mathcal{M}_i$. We extend the derivation ∂ on k to the fields $\tilde{\mathcal{M}}$, $\mathcal{M}_i$. We also choose continuous paths γ_i's joining x_0, respectively, to the d_i's (that is, arriving at a_i tangentially to d_i).*

Let ∇ be a meromorphic connection on X with poles on S at most. We denote by ∇_i the germ at a_i of ∇. There exists a uniquely determined Picard-Vessiot extension L_0 (resp., L_i) of the differential field $(K; \partial)$ (resp., $(K_i; \partial)$) associated to ∇ (resp., ∇_i) such that $K \subset L_0 \subset \tilde{\mathcal{M}}$ (resp., $K_i \subset L_i \subset \mathcal{M}_i$). The path γ_i induces an isomorphism of differential fields Z_i between L_0 and L_i (we use Cauchy's theorem and analytical extension along γ_i).

We denote by G (resp., G_i) the "representation" of "the" differential Galois group $\mathrm{Gal}_K \nabla$ (resp., $\mathrm{Gal}_{K_i} \nabla$) associated to L_0 (resp., L_i). Using Z_i, we identify the local Galois group G_i with a subgroup of the global Galois group G. Let Π_1 be the (usual) monodromy group of ∇. Then the complex linear algebraic group G is topologically generated by the G_i's ($i \in I$) and Π_1.

Proof. This result is a trivial extension of a classical result due to Marotte ([68], CH. II, H.30). We recall briefly the proof. We denote by H the subgroup of G generated by the G_i's and Π_1. Let $\alpha \in L_0 \subset \tilde{\mathcal{M}}_0$. It defines elements $\alpha_i \in L_i$. If α is invariant by the group H, then α_i is invariant by G_i and Pi_1. Therefore, $\alpha_i \in K_i$: α is uniform around a_i, and corresponds to a germ of meromorphic function at a_i. Finally α is non-ramified on S and meromorphic on X, with poles on X at most. We have proved that the subfield of the Picard-Vessiot extension L_0, fixed by the subgroup $H \subset G$ is K. Then from the differential Galois correspondence (Theorem 2.2) it follows that H is Zariski dense in G.□

We remark that from Theorem B.1 and from Proposition B.1 it follows a stronger result:

Corollary B.1 *The Galois group G of the connection ∇ in Theorem B.1 is topologically generated by the Stokes matrices, the exponential tori (at the singular points) and the monodromy group Π_1.*

Theorem B.2 *Let X be a connected Riemann surface. Let $f : X \longrightarrow X$ be a finite ramified covering of X by a connected Riemann surface X'. Let ∇ be a meromorphic connection on X. We set $\nabla' = f^*\nabla$. Then we have a natural injective homomorphism*

$$\text{Gal}\,(\nabla') \to \text{Gal}\,(\nabla)$$

of differential Galois groups which induces an isomorphism *between their Lie algebras.*

Proof. Let K (resp., K') be the meromorphic functions field of X (resp., X'). The finite covering $f : X' \longrightarrow X$ is ramified over a finite set $\Sigma \subset X$. Let $S \subset X$ be the union of the ramification set Σ and of the set of poles of ∇. It is a discrete subset. Let $S' = f^{-1}(S) \subset X'$. It is a discrete subset. We choose a base point $x'_0 \in X' - S'$ and we set $f(x'_0) = x_0 \in X$. Then we set $G = \text{Gal}_K(\nabla)$ and $G' = \text{Gal}_{K'}(f^*\nabla)$, with similar conventions to those made above in the proof of Theorem B.1.

The field inclusion $K \subset K'$ induces a natural map

$$\varphi : G' \to G.$$

This map is clearly continuous and injective and we can identify G' with a closed subgroup of G. We have a natural injective map $\pi_1(X' - S'; x'_0) \to \pi_1(X - S; x_0)$. We can identify $\pi_1(X' - S'; x'_0)$ with a subgroup of $\pi_1(X - S; x_0)$. The index of this subgroup is finite. Following our conventions, we compute G (resp., G') with the horizontal sections of ∇ meromorphic on the universal covering pointed at x_0 (resp., x'_0).

We denote by Π_1 (resp., Π_1') the natural image of $\pi_1(X - S; x_0)$ (resp., $\pi_1(X' - S'; x_0')$) in G (resp., G'). By Theorem B.1, the global differential Galois group G (resp., G') is topologically generated by G_i's and Π_1 (resp., $G_{i'}'$ and Π_1'). Let R (resp. R') be the smallest subgroup of G (resp. G') such that it contains the identity components of all the local differential Galois groups, and such that it is invariant by the adjoint action of the monodromy subgroup Π_1 (resp. Π_1'). The group R (resp. R') is closed and connected.

Using Proposition B.1, we see that the group G (resp., G') is topologically generated by Π_1 and R (resp., Π_1' and R'). We choose continuous paths γ_i joining x_0 to each point $a_i \in S$ in $X - S$. Afterwards, for each $a_{i'}'$ above a_i, we choose a continuous path $\gamma_{i'}'$, joining x_0' to $a_{i'}'$ in $X' - S'$. We can suppose that $\gamma_{i'}'$ is a path above γ_i followed by a path above a loop at a_i.

Applying Lemma B.1 and the definition of the fundamental group as a quotient of a set of loops, we can easily see that the map φ induces an isomorphism between R' and R. Therefore the natural map

$$\pi_1(X - S; x_0)/\pi_1(X' - S'; x_0') \to G/G'$$

is Zariski dense (the group G is topologically generated by Π_1 and R). As the first group is finite, it follows that the map is onto, and that the group G/G' is also finite. Therefore, $G^0 = G'^0$.   □

As remarked in Chapter 2, in [51] there is an algebraic version of Theorem B.2. It is possible to transpose Katz's Tannakian argument to the analytic situation. Then we get an injective homomorphism

$$G_{K'}' \to G_K \otimes_K K'$$

inducing an isomorphism of K'-Lie algebras

$$\mathcal{G}_{K'}' \to \mathcal{G}_K \otimes_K K'.$$

But this isomorphism comes by tensorisation $\otimes_{\mathbf{C}} K'$ from a $\mathbf{C}$-linear natural map

$$\text{Lie}\, \varphi : \mathcal{G}_{\mathbf{C}}' \to \mathcal{G}_{\mathbf{C}}.$$

Therefore, φ is an isomorphism of complex Lie algebras. This gives another proof of Theorem B.2.

Appendix C

Connections with Structure Group

In this appendix we will prove an important result of Kolchin. As a corollary we get another proof of the fact that the Galois group of a symplectic meromorphic connection is contained in the symplectic group. As in the other appendices we will follow [77].

Let X be a Riemann surface. We denote by $\mathcal{O}_X$ (resp., $\mathcal{M}_X$) its sheaf of holomorphic (resp., meromorphic) functions.

Let $G \subset GL(n; \mathbf{C})$ be a Zariski connected complex linear algebraic group. We denote by $\mathcal{G} \subset \text{End}(n; \mathbf{C})$ its Lie algebra. As in Appendix A, we denote by $\mathbf{G}^{an}$ (resp., $\mathbf{G}^{me}$) the sheaf of holomorphic (resp., meromorphic) matrix functions whose values belong to $\mathcal{O}_X$ (resp., $\mathcal{M}_X$). We adopt similar notations for functions whose values belong to the Lie algebra $\mathcal{G}$.

We recall that we have already defined a holomorphic G-bundle over X as a holomorphic vector bundle over X, admitting G as a structure group. It is characterized by an element of $H^1(M; \mathbf{G}^{an})$. We have a notion of local G-trivialization of a G-bundle. We also introduced the notion of meromorphic G-bundle (see Appendix A).

Let ∇ be a meromorphic connection on a G-bundle V. Using a local coordinate t and a frame corresponding to a local G-trivialisation, we get a differential operator $\nabla_{\frac{d}{dt}} = \frac{d}{dt} - A$, where A is a meromorphic matrix. If the values of A belong to the Lie algebra $\mathcal{G}$, we say that ∇ is a meromorphic connection with structure group G (or a G-connection) on the G-bundle V. This definition is *independent* of the choice of a trivialization: if the values of a meromorphic invertible matrix P belong to the group G, then the values of the meromorphic matrix $P^{-1}AP - P^{-1}\frac{d}{dt}A$ belong clearly to the Lie algebra $\mathcal{G}$.

Theorem C.1 *Let ∇ be a G-meromorphic connection on a trivial G-bundle V over a connected Riemann surface X. Then its differential Galois group "is" a closed subgroup of G.*

J. J. Morales Ruiz, *Differential Galois Theory and Non-Integrability of Hamiltonian*, Systems, Modern Birkhäuser Classics,
DOI: 10.1007/978-3-0348-0723-4, © Springer Basel 1999

Proof. This result is due to Kolchin, who introduced the notion of G-primitive extension [54]. We will give here a very simple Tannakian proof. Following Chevalley's theorem ([96], 5.1.3. Theorem, page 131), the linear algebraic group $G \subset GL(n; \mathbf{C})$ is the subgroup of $GL(n; \mathbf{C})$ leaving invariant a complex line W_0' in some construction W_0 on the complex vector space $V_0 = \mathbf{C}^n$. To the natural operation of the group G on the construction W_0, there corresponds a natural operation of the Lie algebra $\mathcal{G}$ on the same construction W_0. Clearly, this action also leaves invariant the complex line W_0'. We denote by $G_{W_0} \subset GL(W_0)$ the natural representation of G, and by $\mathcal{G}_{W_0} \subset \mathrm{End}(W_0)$ the corresponding Lie algebra. If we choose a basis of the complex vector space W_0 such that its first vector generates W_0' over $\mathbf{C}$, then the Lie algebra $\mathcal{G}_{W_0}$ corresponds to the Lie algebra of the matrices whose entries belonging to the first column are zero, except perhaps the first one.

To the construction W_0 on V_0 there corresponds a holomorphic vector bundle W. We obtain W from the holomorphic vector bundle V by a similar construction. To the meromorphic connection ∇ on V corresponds similarly a meromorphic connection ∇_W on W. To the complex line W_0' corresponds a trivial sub-line bundle W' of W. We choose a (meromorphic) uniformizing variable over X and a frame of the *trivial G_{W_0}-bundle* W, such that its first element generates the sub-line bundle W'. Then the G_{W_0}-meromorphic connection ∇_W can be interpreted as a system $\frac{d}{dt} - B$, where the meromorphic matrix B takes its values into the Lie algebra $\mathcal{G}_{W_0}$. Consequently, the entries of B that belong to the first column are identically zero, except perhaps the first one. Therefore, the action of the meromorphic matrix B on the sheaf of meromorphic sections of the vector bundle W leaves invariant the subsheaf of meromorphic sections whose values belong to the subbundle W'. Going back to connections, we see that the meromorphic connection ∇_W leaves invariant the subbundle W', and consequently, it induces a *subconnection* $(W', \nabla_{W'}) \subset (W, \nabla_W)$.

By the Tannakian approach to the differential Galois theory (Section 2.4), the differential Galois group H of ∇ is *defined* by the *list* of all the subspaces of all the constructions $C(V_0)$ on the vector space V_0 corresponding to the fibres (in fibre functor sense) of the underlying vector bundles of all the subconnections of the similar connections $\nabla_{C(V)}$ on the similar constructions $C(V)$. But (W_0', W_0) *belongs to this list*, therefore H is a *closed subgroup* of the algebraic group G (which itself can be *defined* by the *only pair* (W_0', W_0)). $\square$

In our monograph we need only the following result corresponding to $G = Sp(n; \mathbf{C})$. (Using Appendix A, if V is a meromorphic symplectic bundle over X, we can suppose that it is a *trivial* symplectic meromorphic bundle.)

Corollary C.1 *Let ∇ be a symplectic meromorphic connection on a meromorphic symplectic bundle V over a connected Riemann surface X. Then its differential Galois group is a closed subgroup of the symplectic group.*

Bibliography

[1] M. Adler, P. van Moerbeke, Linearization of Hamiltonian Systems, Jacobi Varieties and Representation Theory, *Advances in Math.*, **38** (1980), 318–379.

[2] M. Adler, P. van Moerbeke, The complex geometry of the Kovalewski-Painlevé analysis, *Invent. Math.*, **97** (1989), 3–51.

[3] R. Abraham, J.E. Marsden, *Foundations of Mechanics.* Benjamin, London, 1978.

[4] V.I. Arnold, *Mathematical methods in classical mechanics.* Springer-Verlag, Berlin, 1978.

[5] V.I. Arnold, A.L. Krylov, Uniform distribution of points on a sphere and some ergodic properties of solutions of linear ordinary differential equations in a complex region, *Dokl. Akad. Nauk*, **148** (1963), 9–12 (= *Soviet Math. Dokl.*).

[6] E. Artin, *Algèbre géométrique.* Gauthier-Villars, Paris, 1978.

[7] F. Baldassarri, On algebraic solutions of Lamé's differential equation, *J. of Diff. Eq.* **41**, (1981), 44–58.

[8] A. Baider, R.C. Churchill, D.L. Rod, M.F. Singer, On the infinitesimal Geometry of Integrable Systems. *Fields Institute Communications*, **7**, American Mathematical Society, Providence, Rhode Island, 1996.

[9] A. Baider, R.C. Churchill, D.L. Rod, Monodromy and non-integrability in complex Hamiltonian systems, *J. Dynamics Differential Equations*, **2** (1990), 451–481.

[10] D.G. Babitt, V.S. Varadarajan, *Local Moduli for Meromorphic Differential Equations*, Astérisque **169–170**, 1989.

[11] B.K. Berger, D. Garfinkel, E. Strasser, New algorithm for Mixmaster dynamics, *Classical and Quantum Gravity* **14** (1997), L29–L36.

[12] F. Beukers, Differential Galois Theory, *From Number Theory to Physics* W. Waldschmidt, P. Moussa, J.-M. Luck, C. Itzykson Ed., Springer-Verlag, Berlin 1995, 413–439.

J. J. Morales Ruiz, *Differential Galois Theory and Non-Integrability of Hamiltonian*, Systems, Modern Birkhäuser Classics, DOI: 10.1007/978-3-0348-0723-4, © Springer Basel 1999

[13] O.I. Bogoyavlensky, *Qualitative Theory of Dynamical Systems in Astrophysics and Gas Dynamics.* Springer, Berlin, 1985.

[14] A. Borel, *Linear algebraic groups.* Springer-Verlag, New York, 1991.

[15] A. Buium, *Differential Function Fields and Moduli of Algebraic Varieties,* Lectures Notes in Math. 1226. Springer-Verlag, Berlin 1986.

[16] J. Cano, J.P. Ramis, *Théorie de Galois différentielle, multisommabilité et phénomènes de Stokes.* To appear.

[17] H. Cartan, *Espaces fibrés analytiques, d'après H. Grauert.* Séminaire Bourbaki, Décembre 1956, 137-01-12.

[18] R. Carter, *Simple Groups of Lie type.* Wiley, New York, 1972.

[19] R. Carter, G. Seagal, I. Macdonald, *Lectures on Lie Groups and Lie Algebras.* Cambridge University Press, Cambridge, UK, 1995.

[20] J.J.Cornish, J.J.Levin. *Preprint* gr-qc/9605029

[21] R.C. Churchill, A comparison of the Kolchin and Deligne-Katz Definitions of a Differential Galois Group. *The Kolchin Seminar in Differential Algebra,* City College of New York, February, 1998.

[22] R.C. Churchill, Galoisian Obstructions to the Integrability of Hamiltonian Systems. *The Kolchin Seminar in Differential Algebra,* City College of New York, May, 1998.

[23] R.C. Churchill, J. Delgado, D.L. Rod, The Spring-Pendulum System and the Riemann Equation. *New Trends for Hamiltonian Systems and Celestial Mechanics,* E.A. Lacomba, J. Llibre, Ed., World Scientific, Singapore, 1993.

[24] R.C. Churchill, D.L. Rod, Geometrical aspects of Ziglin's Nonintegrability theorem for Complex Hamiltonian Systems, *J. of Diff. Eq.* **76** (1988), 91–114.

[25] R.C. Churchill, D.L. Rod, On the determination of Ziglin monodromy Groups, *SIAM J. Math. Anal.* **22** (1991), 1790–1802.

[26] R.C. Churchill, D.L. Rod, M.F. Singer, Group Theoretic Obstructions to Integrability, *Ergod. Th. and Dynam. Sys* **15**(1995), 15–48.

[27] R.C. Churchill, D.L. Rod, B.D. Sleeman, Linear Differential Equations with Symmetries. in *Ordinary and Partial Differential Equations: Volume 5,* P.O. Smith and R.J. Jarvis, Ed., Pitman Research Notes in Mathematics Series, Addison Wesley-Logmann, Reading, MA, 1997, 108–129.

[28] R. Cushman, J. Sniatycki, Local Integrability of the Mixmaster Model, *Reports on Mathematical Physics* **36** (1995), 75–89.

[29] P. Deligne, Catégories tannakiennes. *The Grothendieck Festschrift, vol. II.* Progress in Math. **87**. Birkhäuser, Boston, 1990, 111–196.

[30] P. Deligne, *Equations Différentielles à points singuliers réguliers.* Lectures Notes in Math.**163**. Springer-Verlag, Berlin, 1970.

[31] J. Dieudonné, *Sur les groupes classiques.* Hermann, Paris 1948.

[32] A. Duval, The Kovacic Algorithm with applications to special functions. *Differential Equations and Computer Algebra*, M. Singer, Ed., Academic Press, London, 1991, 113–130.

[33] A. Duval, M. Loday-Richaud, *Applicable Algebra in Engineering, Communication and Computing* **3** (1992), 211–246.

[34] B. Dwork, Differential operators with nilpotent p-curvature, *Am. J. Math.* **112** (1990), 749–786.

[35] B. Dwork, private communication.

[36] O. Foster, *Lectures on Riemann Surfaces.* Springer-Verlag, New York, 1981.

[37] A. Goriely, M. Tabor, The singularity analysis for nearly integrable systems: homoclinic intersections and local multivaluedness, *Physica D* **85** (1995), 93–125.

[38] B. Grammaticos, B. Dorizi, A. Ramani, Integrability of Hamiltonians with third and fourth degree polynomials potentials, *J. Math. Phys.* **24**(1983), 2289–2295.

[39] H. Grauert, *Approximationssätze und analytische Faserräume*, Math. Annalen **133** (1957), 139–159.

[40] C. Grotta-Ragazzo, Nonintegrability of some Hamiltonian Systems, Scattering and Analytic Continuation, *Commun. Math. Phys.* **166** (1994), 255–277.

[41] A. Haefliger, Local Theory of Meromorphic Connections in dimension one (Fuchs theory). *Algebraic D-modules*, A. Borel, Ed., Academic Press, Boston 1987.

[42] G. H. Halphen, *Traité des fonctions elliptiques* Vol. I, II. Gauthier-Villars, Paris, 1888.

[43] G. Heilbronn, *Intégration des équations différentielles par la méthode de Drach*, Memorial des Sciences mathématiques, Fasciscule CXXXIII. Gauthier-Villars, Paris, 1956.

[44] E. Horozov, On the non-integrability of the Gross-Neweu models, *Annals of Physics* **174** (1987), 430–441.

[45] J.E. Humphreys, *Linear Algebraic Groups*, Springer-Verlag, New York, 1981.

[46] E.L. Ince, *Ordinary differential equations.* Dover, New York, 1956.

[47] M. Irigoyen, C. Simó, Non-integrability of the J2 Problem, *Cel. Mech. and Dynam. Astr.* **55** (1993), 281–287.

[48] H. Ito, Non-integrability of the Hénon-Heiles system and a theorem of Ziglin, *Koday Math. J.* **8** (1985), 129–138.

[49] E. Juillard Tosel, Un résultat de non-intégrabilité pour le potentiel en $1/r^2$, C. R. Acad. Sci. Paris, **t. 327**, Série I (1998), 387–329.

[50] I. Kaplansky, *An Introduction to Differential Algebra*. Hermann, Paris 1976.

[51] N.M. Katz, A conjecture in the arithmetic theory of differential equations. *Bull. Soc. Math. France* **110** (1982), 203–239.

[52] T. Kimura, On Riemann's Equations which are Solvable by Quadratures. *Funkcialaj Ekvacioj* **12** (1969), 269–281.

[53] F. Kirwan, *Complex Algebraic Curves*. Cambridge Univ. Press, Cambridge, UK, 1992.

[54] E. Kolchin, *Differential algebra and algebraic groups*. Academic Press, New York, 1973.

[55] O.Yu. Kolsova, L.M. Lerman, Families of transverse Poincaré homoclinic orbits in $2N$-dimensional Hamiltonian systems close to the system with a loop to a saddle-center, *International Journal of Bifurcation and Chaos* **6** (1996), 991–1006.

[56] J.J. Kovacic, An Algorithm for Solving Second Order Linear Homogeneous Differential Equations. *J. Symb. Comput.* **2** (1986), 3–43.

[57] S. Kowalevski, Sur le problème de la rotation d'un corps solide autour d'un point fixe, *Acta Math.* **12** (1889), 177–232.

[58] M. Kummer, A.W. Sáenz, Non-integrability of the Classical Zeemann Hamiltonian, *Commun. Math. Phys.* **162** (1994), 447–465.

[59] M. Kummer, A.W. Sáenz, Non-integrability of the Störmer problem, *Physica D* **86** (1995), 363–372.

[60] L. Landau, E. Lifchitz, *Théorie des champs*. Mir, Moscou 1970.

[61] A. Latifi, M. Musette, R. Conte, The Bianchi IX cosmological model is not integrable, *Physics Letters A* **194** (1994), 83–92.

[62] N.N. Lebeded, *Special Functions and Their Applications*. Dover, New York, 1972.

[63] E. Leimanis, *The General Problem of the Motion of Coupled Rigid Bodies about a Fixed Point.* Springer Tracts in Natural Philosophy Vol. 7. Springer-Verlag, Berlin, 1965.

[64] L.M. Lerman, Hamiltonian systems with loops of a separatrix of a saddle-center, *Sel. Math. Sov.* **10** (1991), 297–306.

[65] P. Libermann, C.M. Marle, *Symplectic Geometry and Analytical Mechanics*. D. Reidel Publ. Company, Dordrecht, NL, 1987.

[66] A.M. Lyapounov, On a certain property of the differential equations of the problem of motion of a heavy rigid body having a fixed point, *Soobshch. Kharkov Math. Obshch.*, Ser. 2, **4** (1894), 123–140. Collected Works, Vol. 5, Izdat. Akad. SSSR, Moscow (1954) (in Russian).

[67] B. Malgrange, Regular Connections after Deligne. *Algebraic D-modules*, A.Borel, Ed., Academic Press, Boston, 1987.

[68] F. Marotte, Les équations différentielles linéaires et la théorie des groupes, Ann. Fac. Sciences Univ. de Toulouse (1), 12, 1898.

[69] J. Martinet, J.P. Ramis, Théorie de Galois différentielle et resommation. *Computer Algebra and Differential Equations*, E. Tournier, Ed. Academic Press, London, 1989, 117–214.

[70] J. Martinez-Alfaro, C. Chiralt, Invariant Rotational Curves in the Sitnikov Problem. *Celestial Mechanics and Dynamical Astronomy* **55**, (1992), 351–367.

[71] A. R. Magid, *Lectures on Differential Galois Theory*, American Mathematical Society. Providence, Rhode Island, 1994.

[72] K.R. Meyer, G.R. Hall, *Introduction to Hamiltonian Dynamical Systems and the N-Body Problem*. App. Math. Sci. **90**. Springer-Verlag, New York, 1992.

[73] C.W. Misner, Mixmaster Universe. *Physical Review Letters* **20** (1969), 1071–1074.

[74] C. Mitchi, Differential Galois Groups and G-functions. *Differential Equations and Computer Algebra*, M. Singer, Ed., Academic Press, London, 1991, 149–180.

[75] J.J. Morales-Ruiz, *Técnicas algebraicas para el estudio de la integrabilidad de sistemas hamiltonianos*, Ph.D. Thesis, University of Barcelona, 1989.

[76] J.J. Morales-Ruiz, J.M. Peris, On a Galoisian Approach to the Splitting of Separatrices. *Preprint* 1996.

[77] J.J. Morales-Ruiz, J.P. Ramis, Galoisian Obstructions to Integrability of Hamiltonian Systems. *Preprint* 1997.

[78] J.J. Morales-Ruiz, J.P. Ramis, A Note on the Non-Integrability of some Hamiltonian Systems with a Homogeneous Potential. *Preprint* 1997.

[79] J.J. Morales-Ruiz, J.P. Ramis, Galoisian Obstructions to integrability of Hamiltonian Systems II. *Preprint* 1997.

[80] J.J. Morales-Ruiz, J.P. Ramis, C. Simó, Integrability of Hamiltonian Systems and Differential Galois Groups of Higher Variational Equations. To appear.

[81] J.J. Morales-Ruiz, C. Simó, Picard-Vessiot Theory and Ziglin's Theorem. *J. Diff. Eq.* **107** (1994), 140–162.

[82] J.J. Morales-Ruiz, C. Simó, Non-integrability criteria for hamiltonians in the case of Lamé Normal Variational Equations. *J. Diff. Eq.* **129** 1996, 111–135.

[83] J.Moser, *Stable and Random Motions in Dynamical Systems.* Princeton Univ. Press, Princeton, 1973.

[84] D. Mumford, *Tata Lectures on theta II*, Progress in Mathematics, vol. 43. Birkhäuser, Boston, 1984.

[85] M. Namba, *Geometry of Projective Algebraic Curves.* Marcel Dekker Inc., New York, 1984.

[86] A.L. Onishchik, E.B. Vinberg, *Lie Groups and Algebraic Groups.* Springer-Verlag, New York, 1990.

[87] H. Poincaré, *Les Méthodes Nouvelles de la Mécanique Céleste,* Vol. I. Gauthiers-Villars, Paris, 1892.

[88] E.G.C. Poole, *Introduction to the theory of Linear Differential Equations.* Oxford Univ. Press, London, 1936.

[89] A. Ramani, B. Grammaticos, T. Bountis, The Painlevé property and singularity analysis of integrable and non-integrable systems, *Physics Reports* **180** (1989), 159–245.

[90] J.P. Ramis, About my Theorem on Galois differential groups and Stokes multipliers: A "mode d'emploi". *Preprint*, Université Louis Pasteur. I.R.M.A. Strasbourg, 1993.

[91] J.P. Serre, *Géométrie algébrique et géométrie analytique*, Ann. Inst. Fourier Grenoble, 1956, 1–42.

[92] Y. Sibuya, *Linear Differential Equations in the Complex Domain: Problems of Analytic Continuation.* Transl. of Math. Monogr. 82, Am. Math. Soc. Providence, Rhode Island 1990.

[93] C.L. Siegel, *Topics in complex function theory.* Wiley, New York, 1969.

[94] M.F. Singer, An outline of Differential Galois Theory in *Computer Algebra and Differential Equations*, E. Tournier, Ed., Academic Press, New York, 1989, 3–57.

[95] M.F. Singer, Liouvillian solutions of linear differential equations with Liouvillian coefficients, *J. Symbolic Computation* **11** (1991), 251–273.

[96] T.A. Springer, *Linear Algebraic Groups*, Birkhäuser, Boston, 1981.

[97] M.F. Singer, F. Ulmer, Galois Groups of second and third order linear differential equations, *J. Symbolic Computation* **16** (1993), 1–36.

[98] M.F. Singer, F. Ulmer, Liouvillian and algebraic solutions of second and third order linear differential equations, *J. Symbolic Computation* **16** (1993), 37–73.

[99] A.H. Taub, Empty Space-Times admitting a Three Parameter Group of motions. *Annals of Mathematics* **53** (1951), 472–490.

[100] F. Ulmer, J.A. Weil, Note on Kovacic's Algorithm, *J. Symbolic Computation* **22** (1996), 179–200.

[101] K. Umeno, Non-integrable character of Hamiltonians Systems with global and symmetric coupling, *Physica D* **82** (1995), 11–35.

[102] M. van der Put, *Galois theory of differential equations, Algebraic groups and Lie algebras, Preprint* 1997.

[103] P. Vanhaecke, *Integrable Systems in the realm of Algebraic Geometry.* Lecture Notes in Math. 1638. Springer-Verlag, Berlin, 1996.

[104] V.S. Varadarajan, Meromorphic differential equations. *Expositiones Mathematicae* **9** (1991), 97–188.

[105] W. Wasow, *Asymptotic Expansions for Ordinary Differential Equations.* Dover, New York, 1965.

[106] J.A. Weil, *Constantes et polynômes de Darboux en algèbre différentielle: applications aux systèmes différentiels linéaires.* Ph.D. Thesis, Ecole Polytechnique 1995.

[107] A. Weinstein, Lagrangian submanifolds and Hamiltonian systems, *Ann. of Math.* **98** (1973), 377–410.

[108] H. Weyl, *Meromorphic Functions and Analytic Curves.* Princeton Univ. Press, Princeton, NJ, 1943.

[109] E.T. Whittaker, E.T. Watson, *A Course of Modern Analysis.* Cambridge Univ. Press, Cambridge, UK, 1989.

[110] H. Yoshida, Existence of exponentially unstable periodic solutions and the non-integrability of homogeneous hamiltonian systems, *Physica D***21**, (1986), 163–170.

[111] H. Yoshida, A criterion for the non-existence of an additional integral in hamiltonian systems with a homogeneous potential, *Physica D***29**, (1987), 128–142.

[112] H. Yoshida, Exponential instability of the collision orbits in the anisotropic Kepler problem, *Celestial Mechanics* **40** (1987), 51–66.

[113] H. Yoshida, On a class of variational equations transformable to the Gauss hypergeometric equation, *Celestial Mechanics and Dynamical Astronomy* **53** (1992), 145–150.

[114] S.L. Ziglin, Branching of solutions and non-existence of first integrals in Hamiltonian mechanics I, *Funct. Anal. Appl.* **16** (1982), 181–189.

[115] S.L. Ziglin, Branching of solutions and non-existence of first integrals in Hamiltonian mechanics II. *Funct. Anal. Appl.* **17** (1983), 6–17.

Index

J. J. Morales Ruiz, *Differential Galois Theory and Non-Integrability of Hamiltonian*, Systems, Modern Birkhäuser Classics, DOI: 10.1007/978-3-0348-0723-4, © Springer Basel 1999